I0049487

Manufacturing, Characterisation and Properties of Advanced Nanocomposites

Manufacturing, Characterisation and Properties of Advanced Nanocomposites

Special Issue Editors

Yu Dong
Alokesh Pramanik
Dongyan Liu
Rehan Umer

MDPI • Basel • Beijing • Wuhan • Barcelona • Belgrade

MDPI

Special Issue Editors

Yu Dong
School of Civil and Mechanical
Engineering, Curtin University
Australia

Alokesh Pramanik
School of Civil and Mechanical
Engineering, Curtin University
Australia

Dongyan Liu
Institute of Metal Research
Chinese Academy of Sciences
China

Rehan Umer
Centre for Future Materials
University of Southern Queensland Toowoomba
Australia

Rehan Umer
Department of Aerospace Engineering Khalifa
University of Science and Technology
UAE

Editorial Office
MDPI
St. Alban-Anlage 66
Basel, Switzerland

This is a reprint of articles from the Special Issue published online in the open access journal *Journal of Composites Science* (ISSN 2504-477X) from 2017 to 2018 (available at: http://www.mdpi.com/journal/jcs/special issues/nanocomposites)

For citation purposes, cite each article independently as indicated on the article page online and as indicated below:

LastName, A.A.; LastName, B.B.; LastName, C.C. Article Title. *Journal Name* **Year**, *Article Number, Page Range*.

ISBN 978-3-03897-188-7 (Pbk)
ISBN 978-3-03897-189-4 (PDF)

Articles in this volume are Open Access and distributed under the Creative Commons Attribution (CC BY) license, which allows users to download, copy and build upon published articles even for commercial purposes, as long as the author and publisher are properly credited, which ensures maximum dissemination and a wider impact of our publications. The book taken as a whole is © 2018 MDPI, Basel, Switzerland, distributed under the terms and conditions of the Creative Commons license CC BY-NC-ND (http://creativecommons.org/licenses/by-nc-nd/4.0/).

Contents

The Effect of Polycaprolactone Nanofibers on the Dynamic and Impact Behavior of Glass Fibre
Reinforced Polymer Composites
Reprinted from: *J. Compos. Sci.* **2018**, 2, 43, doi: 10.3390/jcs2030043 . . . . **128**

About the Special Issue Editors

Yu Dong received a BE in Mechanical Engineering and Automation from Wuhan University of Science and Technology (WUST), China in 2000, as well as an ME (1st Class Honours) and a PhD in Mechanical Engineering from the University of Auckland, New Zealand in 2002 and 2008, respectively. He is currently a senior lecturer at the School of Civil and Mechanical Engineering, Curtin University, Australia. His research interests include polymer nanocomposites, electrospun nanofibres, nanomaterial processing and characterisation, micromechanical modelling, additive manufacturing, and statistical design of experiments. He has published two edited books, entitled "Nanostructures: Properties, Production Methods and Applications" by NOVA Science Publishers, USA in 2013 and "Fillers and Reinforcements for Advanced Nanocomposites" by Elsevier, UK in 2015. His other widespread publications include 11 book chapters, 60 peer-review journal articles, and 32 referred conference papers. He serves as an Associate Editor for *Frontiers in Materials* and an Editorial Board Member of *Journal of Composites Science, Heliyon, Journal of Hydromechatronics* and *SpringerPlus*.

Alokesh Pramanik received his BSc in Mechanical Engineering from Bangladesh University of Engineering and Technology, Bangladesh. He then completed his Master and PhD degrees in Mechanical Engineering at the National University of Singapore and the University of Sydney, respectively. He is currently working as a Lecturer in Mechanical Engineering at the School of Civil and Mechanical Engineering, Curtin University, Perth, Australia. His research interests include manufacturing processes, composite materials, and finite element analysis.

Dongyan Liu received her PhD degree from the Institute of Metal Research, Chinese Academy of Sciences, China in 2005. She then worked as a postdoctoral research fellow on electro-spark deposited amorphous coatings at the Department of Chemical and Materials, University of Auckland, New Zealand. She started her research career on polymer nanocomposites at the Centre for Advanced Composite Materials, University of Auckland in 2008. She is currently an Associate Professor at the Institute of Metal Research, Chinese Academy of Sciences, China. Dr. Liu's research interests focus on polymer nanocomposites reinforced with cellulose nanowhiskers and graphene nanosheets. She has authored and co-authored over 40 papers in fully refereed journals as well as three book chapters, and has presented her research at many international conferences.

Rehan Umer received his PhD degree from the University of Auckland, New Zealand in 2008. He then worked as a Research Engineer at CRC-ACS, Brisbane, Australia. He started his postdoctoral research career at the Composite Vehicle Research Centre (CVRC), Michigan State University, USA on polymer composites processing. In 2012, he was hired as an Assistant Professor in the Department of Aerospace Engineering at Khalifa University of Science and Technology, Abu Dhabi, UAE. He is the co-founder of the Aerospace Research and Innovation Centre (ARIC), a 15 million-dollar joint venture between Khalifa University of Science and Technology and Mubadala Aerospace. He later became an Associate Professor and Associate Director of ARIC. Dr. Umer's research has focused on advanced composites manufacturing covering both experimental and modelling studies. Dr. Umer has participated in several industrial collaborations and he has also edited a book, authored and co-authored many papers in refereed journals as well as book chapters, and presented at a number of international conferences.

Preface to "Manufacturing, Characterisation and Properties of Advanced Nanocomposites"

Advanced nanocomposites stem from the use of small fractions of nanofillers with good dispersion to generate remarkable property enhancement as well as favourable cost-effectiveness, including but not limited to better mechanical, thermal, electrical, and barrier properties. The nanofillers include carbon nanotubes (CNTs), nanoclays, graphene oxides (GOs), cellulose nanowhiskers, metal nanoparticles, and electrospun nanofibres. The typical merit of nanocomposites lies in their easy processability and their ability to rely on current manufacturing technologies like melt compounding, solution casting, in situ polymerisation, and electrospinning. Enormous commercial opportunities for advanced nanocomposites have arisen in the automobile, aerospace and aerocraft, building structures, and biomedical devices over the last decade.

Well-tailored advanced nanocomposites targeting end-users' applications are often desirable to material engineers and manufacturers, depending primarily on a typical good processing-structure-property relationship. For instance, significantly improved mechanical properties of nanocomposites are most likely to result from uniform nanofiller dispersion in morphological structures, which can directly benefit from an initial optimum material manufacturing process with the aid of high shear mixing or ultrasonication to substantially reduce the material size from large aggregates, microfillers to nanofillers. Similar cases also apply as far as thermal and barrier properties of nanocomposites are concerned. As such, the main focus of this edited book is to showcase the latest progress in manufacturing, characterisation and properties of advanced nanocomposites in order to provide technical guidance to this new composite family with the incorporation of nanoparticles and nanomaterials. This book can also be a good reference to material scientists and industrialists working with composites, as well as researchers and engineers dealing with nanotechnology and nanomaterials, leading to the ultimate achievement of a "bottom-up" scheme.

The editors would like to use this opportunity to thank all of the authors who contributed research papers to this book. We are also indebted to the dedicated reviewers who gave invaluable and constructive comments and feedback, greatly improving the paper quality. Finally, a special acknowledgement goes to the MDPI editorial team at *Journal of Composites Science* including Ms. Wenwei Li, Mr. Loker He, and Ms. Wendy Yang for their kind assistance in the book production.

Yu Dong, Alokesh Pramanik, Dongyan Liu, Rehan Umer
Special Issue Editors

Journal of
composites science

MDPI

Editorial

Manufacturing, Characterisation and Properties of Advanced Nanocomposites

Yu Dong [1,*], Alokesh Pramanik [1], Dongyan Liu [2] and Rehan Umer [3,4]

[1] School of Civil and Mechanical Engineering, Curtin University, GPO Box U1987, Perth, WA 6845, Australia; alokesh.pramanik@curtin.edu.au

[2] Titanium Alloys Division, Institute of Metal Research (IMR), Chinese Academy of Sciences (CAS), Shenyang 110016, China; dyliu@imr.ac.cn

[3] Centre for Future Materials, University of Southern Queensland, Toowoomba, QLD 4350, Australia; rehan.umer@usq.edu.au

[4] Department of Aerospace Engineering, Khalifa University of Science and Technology, Abu Dhabi 127788, UAE

* Correspondence: Y.Dong@curtin.edu.au; Tel.: +61-8-9266-9055

Received: 1 August 2018; Accepted: 1 August 2018; Published: 6 August 2018

Advanced nanocomposites have demonstrated great potential over conventional composites owing to the incorporation of well-dispersed nanofillers with extremely small sizes of a few hundred nanometers in platelet-like, fibre-like, tubular, and spherical shapes. The enormous commercial success of nanocomposites, such as the use of polypropylene/clay nanocomposites to manufacture General Motors' step-assist, sail panels, centre bridges, and box-rail protectors, reveals the material innovation and development in this field. The breakthroughs in advanced nanocomposites, from the research and development viewpoint, rely primarily on how to achieve homogeneous nanofiller dispersion within matrices by means of mechanical processing with high shear stress or chemical modification or treatment. As such, a good understanding of the processing–structure–property relationship is critical to design robust and well-tailored nanocomposites for their wide range of applications.

This special issue consists of 10 research papers covering advanced nanocomposites reinforced with nanoclays, cellulose nanowhiskers, graphene oxides (GOs), metal nanoparticles, electrospun nanofibres, etc., in relation to their manufacturing, characterisation, and properties. Wang et al. [1] investigated the effect of polyhedral oligomeric silsesquioxane (POSS) and hydroxyapatite (HA) nanoparticles on reinforcing chitosan (CS) in biocomposite fibres. Their results revealed that only fracture-related properties became sensitive to the effects resulting from the interaction between nanoparticle type and concentration. Rahman and Wu [2] carried out a holistic computational study to evaluate the effects of processing parameters, such as the extent of clay exfoliation and clay volume fraction, on the nonlinear elastoplastic behaviour of polymer/nanoclay composites, which was based on the implementation of representative volume element (RVE) in finite element analysis (FEA). It was found that large aspect ratios of clay platelets with full clay exfoliation was crucial for achieving preferable mechanical properties of nanocomposites reinforced with nanoclays. This work also offered guidelines for the computer-aided design of processing-property-tailored nanocomposites. Liu et al. [3] electrospun polylactic acid (PLA)/cellulose nanowhisker (CNW) composite nanofibres and confirmed uniform CNW distribution within PLA nanofibres along the direction of the fibre axis. Besides, the water absorption of PLA nanofibres were effectively improved with embedded CNWs. It is anticipated that both CNWs and their composite nanofibres can lead to widespread applications for biomedical engineering, sensors, and nanofiltration. Pramanik et al. [4] utilised a single insert milling tool to assess the face milling of nanoparticles reinforced Al-based metal matrix composites (nano-MMCs). The impacts of feed and speed on machined surfaces of nano-MMCs in relation to surface roughness, profile and appearance, chip surface and ratio, machining forces, and force signals were analysed in a systematic manner. Umer [5] studied the processing characteristics and mechanical properties of glass fabric reinforcements coated with graphene nanoparticles in epoxy composites via vacuum-assisted resin transfer moulding (VARTM). The relevant results indicated that flexural strengths of composites decreased with increasing the weight fraction of graphene nanoparticles despite no change in flexural modulus. On the other hand, the ply-delamination phenomenon occurred arising from the coating of graphene nanoparticles on glass reinforcements, which also generated localised damage resistance under low-velocity impact as opposed to pure glass samples. Chen et al. [6] implemented a sol–gel method to prepare epoxy/multilayer GO composites. It was evidently shown that both the thermal stability and flame retardancy of composites could be improved with the addition of GOs with modified silicon. Rao et al. [7] worked on the effect of electrical conductivity reduction of GOs as effective nanofillers in thin films by using a partial factorial design of experiments based on the Taguchi method. The experimental findings suggested that the electrical resistivity of GOs highly depended on the type of acid treatment, and that samples treated with hydroiodic acid had the lowest resistivity of ~0.003 $\Omega \cdot$cm. Basak et al. [8] examined the property improvement of Sn–Ag–Cu (SAC) alloys by incorporating two different types of Fe and Al_2O_3 nanoparticles. The addition of Fe nanoparticles led to the formation of $FeSn_2$ intermetallic compounds (IMCs) along with Ag_3Sn and Cu_6Sn_5 from monolithic SAC alloys, while Al_2O_3 nanoparticles acted as a grain refiner with good dispersion along primary β-Sn grain boundaries without the contribution to phase formation. Notwithstanding that insignificant effect arose from the inclusion of Fe and Al_2O_3 on the thermal behaviour of nanocomposites, their nanoreinforcement potentially gave rise to better mechanical performance when compared with that of conventional monolithic SAC solder alloys. Nakagaito et al. [9] explored reinforcing PLA by the combination of cellulose and chitin nanofibres rather than a single reinforcement phase. Such nanocomposites demonstrated higher tensile properties than those reinforced with cellulose or chitin nanofibres alone. It appeared that chitin acted as a compatibiliser between hydrophobic PLA and hydrophilic cellulose, which played a complementary role along with cellulose in view of the formation of a rigid cellulose nanofibre percolated network. Finally, Garcia et al. [10] focused on understanding the effect of polycaprolactone nanofibres on the dynamics and impact behaviour of polymer/glass fibre composites, in which a finite element model was employed to simulate their impact effect. The numerical results coincided with previous experimental data to prove that composites reinforced with polycaprolactone nanofibres possessed more damage resistance when subjected to the same impact as pristine composites. More importantly, interleaving with polycaprolactone nanofibres was revealed to control the vibrations and improve the resistance of impact damage to structures made of composite mats, which could be used for aircrafts or wind turbines.

All editors would also like to acknowledge the fine contributions of all the authors for their paper submissions to this special issue, as well as the dedicated reviewers for providing timely comments/feedback.

Conflicts of Interest: The authors declare no conflict of interest.

References

1. Wang, K.; Pasbakhsh, P.; De Silva, R.T.; Goh, K.L. A Comparative Analysis of the Reinforcing Efficiency of Silsesquioxane Nanoparticles versus Apatite Nanoparticles in Chitosan Biocomposite Fibres. *J. Compos. Sci.* **2017**, *1*, 9. [CrossRef]
2. Rahman, A.; Wu, X.F. Computational Study of the Effects of Processing Parameters on the Nonlinear Elastoplastic Behavior of Polymer Nanoclay Composites. *J. Compos. Sci.* **2017**, *1*, 16. [CrossRef]
3. Liu, W.; Dong, Y.; Liu, D.; Bai, Y.; Lu, X. Polylactic Acid (PLA)/Cellulose Nanowhiskers (CNWs) Composite Nanofibers: Microstructural and Properties Analysis. *J. Compos. Sci.* **2018**, *2*, 4. [CrossRef]
4. Pramanik, A.; Basak, A.K.; Dong, Y.; Shankar, S.; Guy, L. Milling of Nanoparticles Reinforced Al-Based Metal Matrix Composites. *J. Compos. Sci.* **2018**, *2*, 13. [CrossRef]

5. Umer, R. Manufacturing and Mechanical Properties of Graphene Coated Glass Fabric and Epoxy Composites. *J. Compos. Sci.* **2018**, *2*, 17. [CrossRef]

6. Chen, M.H.; Ke, C.Y.; Chiang, C.L. Preparation and Performance of Ecofriendly Epoxy/Multilayer Graphene Oxide Composites with Flame-Retardant Functional Groups. *J. Compos. Sci.* **2018**, *2*, 18. [CrossRef]

7. Rao, S.; Upadhyay, J.; Polychronopoulou, K.; Umer, R. Reduced Graphene Oxide: Effect of Reduction on Electrical Conductivity. *J. Compos. Sci.* **2018**, *2*, 25. [CrossRef]

8. Basak, A.K.; Pramanik, A.; Riazi, H.; Silakhori, M.; Netting, K.O. Development of Pb-Free Nanocomposite Solder Alloys. *J. Compos. Sci.* **2018**, *2*, 28. [CrossRef]

9. Nakagaito, A.N.; Kanzawa, S.; Takagi, H. Polylactic Acid Reinforced with Mixed Cellulose and Chitin Nanofibers-Effect of Mixture Ratio on the Mechanical Properties of Composites. *J. Compos. Sci.* **2018**, *2*, 36. [CrossRef]

10. Garcia, C.; Trendafilova, I.; Zucchelli, A. The Effect of Polycaprolactone Nanofibers on the Dynamic and Impact Behavior of Glass Fibre Reinforced Polymer Composites. *J. Compos. Sci.* **2018**, *2*, 43. [CrossRef]

© 2018 by the authors. Licensee MDPI, Basel, Switzerland. This article is an open access article distributed under the terms and conditions of the Creative Commons Attribution (CC BY) license (http://creativecommons.org/licenses/by/4.0/).

Journal of
composites science

MDPI

Article

A Comparative Analysis of the Reinforcing Efficiency of Silsesquioxane Nanoparticles versus Apatite Nanoparticles in Chitosan Biocomposite Fibres

Kean Wang [1], Pooria Pasbakhsh [2], Rangika Thilan De Silva [3] and Kheng Lim Goh [4,5,*]

[1]   Department of Chemical Engineering, The Petroleum Institute, Abu Dhabi, P.O. Box 2533, UAE;
      kwang@pi.ac.ae
[2]   School of Engineering, Monash University Malaysia, Level 4, Building 5, Jalan Lagoon Selatan,
      Bandar Sunway, Selangor Darul Ehsan, Subang Jaya 47500, Malaysia; pooria.pasbakhsh@monash.edu
[3]   SLINTEC (Pvt) Limited, Sri Lanka Institute of Nanotechnology, Nanotechnology & Science Park,
      Mahenwatte, Pitipana, Homagama, Sri Lanka; rangikads@slintec.lk
[4]   Faculty of Science, Agriculture and Engineering, Newcastle University, Devonshire Building,
      Newcastle NE1 7RU, UK
[5]   Newcastle University Singapore, SIT Building @ Nanyang Polytechnic, 172A Ang Mo Kio Avenue 8, #05-01,
      Singapore 567739, Singapore
*   Correspondence: kheng-lim.goh@ncl.ac.uk; Tel.: +65-9757-8847

Received: 23 June 2017; Accepted: 14 August 2017; Published: 18 August 2017

Abstract: A comparative analysis of the effects of polyhedral oligomeric silsesquioxane (POSS) and hydroxyapatite (HA) for reinforcing chitosan (CS) is given here. Wet-spun CS nanocomposite fibres, blended with HA or POSS nanoparticles, at varying concentrations ranging from 1 to 9% (w/w) were stretched until rupture to determine the mechanical properties related to the elasticity (yield strength and strain, stiffness, resilience energy) and fracture (fracture strength strain and toughness) of the composite. Two-factor analysis of variance of the data concluded that only the fracture-related properties were sensitive to interaction effects between the particle type and concentration. When particle type is considered, the stiffness and yield strength of CS/POSS fibres are higher than CS/HA fibres—the converse holds for yield strain, extensibility and fracture toughness. With regards to sensitivity to particle concentration, stiffness and yield strength reveal trending increase to a peak value (the optimal particle concentration associated with the critical aggregation) and trending decrease thereafter, with increasing particle concentration. Although fracture strength, strain at fracture and fracture toughness are also sensitive to particle concentration, no apparent trending increase/decrease is sustained over the particle concentration range investigated here. This simple study provides further understanding into the mechanics of particle-reinforced composites—the insights derived here concerning the optimized mechanical properties of chitosan composite fibre may be further developed to permit us to tune the mechanical properties to suit the biomedical engineering application.

Keywords: hydroxyapatite; polyhedral oligomeric silsesquioxanes; elasticity; fracture; particle shape; particle concentration; Weibull model

---

1. Introduction

To overcome the limitations of biopolymers, such as low stiffness and strength, and to enable these materials to have wide applicability, inorganic particulate fillers are often blended with the biopolymer to form a composite material that possesses enhanced stiffness and strength [1–3]. This study is concerned with chitosan (CS) biopolymer, a linear polysaccharide that can be derived from the alkaline N-deacetylation of chitin [4,5]. Chitin is the main structural component of the shells of crustaceans, the exoskeletons of insects and the cell walls of fungi [4–8]. CS polysaccharide is a copolymer comprising β-(1-4)-2-amino-D-glucose (deacetylated unit) and β-(1-4)-2-acetamido-D-glucose (acetylated unit) [4,5]. When the amount of deacetylated unit is higher than 50%, the resultant compound is known as chitosan, otherwise it is known as chitin [4,5]. CS has been a subject of great interest owing to its biocompatibility-related advantages, notably anti-clotting, biodegradability [4,9], antimicrobial and low toxicity (even in blends) [10–12]. Thus, it can be used for medical products such as bandages and implants, or grafted with compounds to yield chelating agents that can be used in water filters [10–12].

A key concern about employing CS in a number of these applications is the low mechanical strength and stiffness [5]. The low values of the mechanical properties of CS limit the applicability for tissue engineering, particularly as implants for soft connective tissues. According to Di Martino et al. [13] and the references therein, the hydrated CS has tensile moduli of 0.1–0.5 MPa (porous membrane) but 5–7 MPa (non-porous membrane). Albanna et al. [14] reported that dehydrated CS fibres have tensile moduli of 2–10 MPa. CS membranes appear to possess higher stiffness; tensile modulus for CS (solution-casted) membranes was found to be in the range of 400–800 MPa [15,16]. Electrospun CS membranes could exhibit a lower tensile modulus of about 300 MPa [17]. CS (solution-casted) membranes have an extensibility (i.e., maximum strain to rupture) of about 0.3 [16]; the extensibility of electrospun membranes varied from 0.3 [13,17] to 1.0 [13]; this variability depends on the pore size and pore orientation [13]. Albanna et al. [14] reported that CS fibres have extensibility values ranging from 0.10 to 0.25. With regards to tensile strength, porous CS structures were found in the range of 30–60 kPa (Di Martino et al. [13] and therein). Liu et al. [15] reported that the CS (solution-casted) membrane has a tensile strength of about 20 MPa. Albanna et al. [14] reported that the CS fibres have tensile strength of 0.4–1.4 MPa. Overall, the strength and modulus of CS materials are much lower than those of soft connective tissue [18], such as skin [19], tendons [20–22] and ligaments [23,24].

To this end, attempts have been carried out to 'tune' the mechanical properties of the CS material by blending with nanoparticles made from, e.g., hydroxyapatite (HA) [25] or silsesquioxane [26,27]. The physical and chemical properties of blends of CS containing HA particles have been well investigated [25,28–34], and CS/HA composites have been proposed for making implantable scaffolds to achieve the desired magnitude of the respective mechanical properties by 'tuning' the HA concentration [30]. The optimised composite is then expected to be useful for influencing the lineage of scaffold-seeded stem cells to generate an extracellular matrix that is compatible with the microenvironment of the host tissue, as well as provide structural support to the host tissue [14]. Blends of CS containing silsesquioxane, namely polyhedral oligomeric silsesquioxane (POSS) particles, have attracted some investigations [26,35,36] and may be a potential alternative to CS/HA. POSSs are compounds having a polyhedral siloxane cage, with the formula $(RSiO_{1.5})_n$ (where $n = 6, 8, 10, 12$ and R = H or organic substituents) [27]. The most common molecular formula of POSSs is $n = 8$; the overall size of the molecule is about 1–3 nm [27]. The key practical advantages of using POSS for blending with polymers are summarized as follows (Reno et al. [37], Blanco et al. [27] and therein). (A) POSS are organosilica three-dimensional, cubic building blocks containing an inorganic inner siloxane core that can be chemically modified at each of the eight corners of the POSS unit. (B) POSS featuring reactive organic groups can be employed as cross-link agents for the preparation of hybrid hydrogel samples. (C) POSS molecules are physically dispersed through weak interactions with the polymeric matrix. This latter approach has important advantages in terms of low cost and synthesis time. (D) It is noted that a homogeneous dispersion of POSS in the matrix may be easily obtained [38–41]. However, only a limited amount of work has been carried out on CS/POSS blends compared to CS/HA blends. For structural applications, the single most important question that has yet to be addressed adequately is how do the magnitudes of the mechanical properties of CS/POSS compare to CS/HA, all things being equal?

To this end, an in-depth analysis of POSS and HA particles for reinforcing CS composite materials has been carried out. Here, both CS/POSS and CS/HA fibres were synthesized by a wet spinning method. The analysis addresses the effects of the particle types (i.e., POSS or HA) and concentration as well as the interaction between the two factors on CS composite. The arguments that underpin the effects of particle concentrations have been investigated for the respective particles and reported in the literature [25,26]. However, most of the reports involve experiments with particle concentration as a single treatment and the experimental conditions are not necessarily the same. Thus, from the material design perspective, optimisation-related arguments concerning the effects of particle types and concentration and the interaction between the two on CS composite seem to be much less well established. To clarify which types of particle would exhibit advantages over the other, as well as to broaden our understanding of the underlying mechanisms of nanoparticulates for reinforcing chitosan composite material, this study investigates the effects of particle type, at varying particle concentrations, on the chitosan composite elasticity and fracture. We hypothesize that the underlying shape of the respective particle type and agglomeration, as well as the interaction between the two factors, influences the mechanical properties of the chitosan composite. A statistical approach, i.e., the two-factor analysis of variance (ANOVA), was used to evaluate the mechanical data and determine the influence of particle type and concentration on the mechanical properties of these CS composite fibres. The mechanical reliability of the nanocomposite was analysed using the Weibull model.

2. Materials and Methods

2.1. Preparation of Chitosan Fibres

The POSS used in this study refers to aminopropylphenyl POSS (AM0272; Hybrid Plastic Inc., Hattiesburg, MS, USA). HA was obtained from Sigma-Aldrich, St. Louis, MO, USA. The CS (85% deacetylated) used in the study were also purchased from Sigma-Aldrich.

CS-based solutions were prepared at the predetermined particle concentrations of the respective HA and POSS, following a protocol that had been reported in a previous study for CS/POSS fibres [26]. Here, a 1.5% w/v concentration of CS to acetic acid was prepared (NB: 1% w/v is equivalent to 0.01 g/mL), and 10 mL of 1% concentration acetic acid was used as a solvent to dissolved 0.15 g of CS. The solution was then stirred at a constant rate of 700 rpm at room temperature. HA and POSS of concentrations 1% (w/w), 3% (w/w), 5% (w/w), 7% (w/w) and 9% (w/w) were added to the CS. (These concentrations corresponded to the masses of 0.0015 g, 0.0045 g, 0.0075 g, 0.0105 g, and 0.0135 g, respectively.) For the CS/POSS blending process, first of all the CS solution was stirred at a constant rate of 700 rpm for 4 h to ensure that the CS was completely dissolved. Consequently, after the POSS was added, the blend was stirred at a constant rate of 700 rpm at room temperature for 18 h before it was introduced into the wet spinning device. Similary, for the CS/HA blending process, the CS solution was stirred at constant rate of 700 rpm for 4 h before the HA was added. After the HA was added, in order to avoid prolonged exposure of HA in an acidic pH environment, the CS/HA blend was stirred at constant rate of 700 rpm for 2 h before it was subjected to the wet spinning process.

CS-based fibres were processed by a wet spinning method, outlined as follows. Figure 1A illustrates a setup of the apparatus for this method. Overall, the process involved (A) extruding the CS solution (containing the particles), through a spinneret, into a coagulation bath containing a non-solvent where the precipitation of the CS-based fibres occurred [25,26,42]; (B) washing the fibres to remove coagulant remains; and (C) winding up the fibres using a bobbin [25,26,42]. The dope (i.e., the blend solution) was made to flow through a syringe pump set at rate 5 mL /min. This dope was then pumped through an epoxy-cured silicone tube into a bath of 1 M of NaOH (i.e., a coagulation bath). Before starting to pump the dope through the system, the tube was first used to deliver a 1 M NaOH solution at a rate of 5 mL/min. Then, the dope was mixed with the coagulant solution at a junction; it was in this bath that the CS would precipitate into a fibre-like form. The CS-based fibres were left in the coagulant solution for 15 min (Figure 1B); thereafter they were removed from the coagulant solution and rinsed with deionised water. Removing the fibres from the solution by winding them around a cylindrical bobbin helped to laterally deform the fibre to achieve a ribbon-like cross section (inset in Figure 1B). The result of the ribbon-like shape resembled a near-triangular cross section when viewed under a field-emission scanning electron microscope (FE-SEM, JEOL JSM-6390LA, JEOL Ltd., Tokyo, Japan) (results not shown).

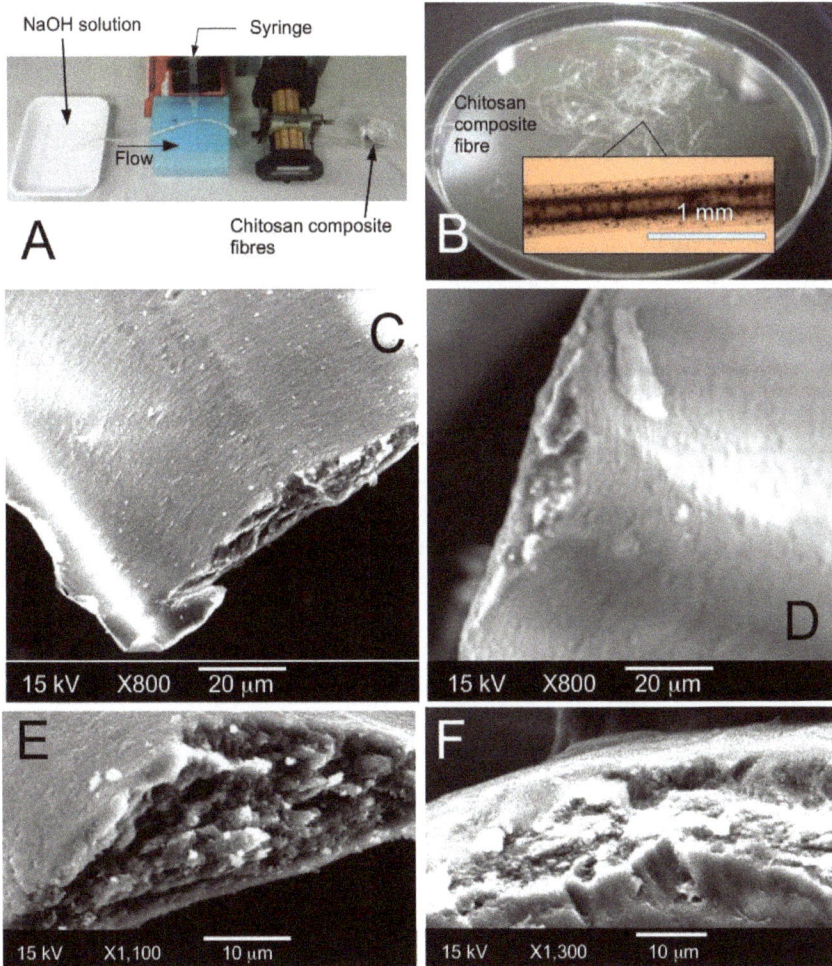

Figure 1. The chitosan composite fibres. (**A**) Set-up of the wet spinning process; (**B**) wet-spun chitosan-based fibres in a petri dish (inset shows a micrograph of the fibre taken using an optical microscope). Low magnification scanning electron micrographs (SEMs) of the cross-section of chitosan fibre reinforced by hydroxyapatite (HA) (**C**) and polyhedral oligomeric silsesquioxane (POSS) (**D**); panels (**E**) and (**F**) show the SEMs of the respective fibres at higher magnification.

2.2. Tensile Testing

The fibres were tested to rupture using a custom-built micromechanical tester [26]. Ten tensile specimens were prepared—according to a method described in previous study [26]—for each of the different combinations of particle type and concentration. Before the test began, the gauge length (i.e., grip-to-grip distance) and cross-sectional area (identified with the area of a triangle) of each fibre specimen were recorded. Specimens were stretched to rupture at a displacement rate of 0.067 mm/s. The fracture morphology of the microstructure was examined using a FE-SEM. The force versus displacement data for the respective specimen were evaluated to derive the stress–strain curve. Here, stress was determined from the force divided by the cross-sectional area of the fibre; strain was determined from the ratio of the fibre displacement to the gauge length. Following the definition based on previous reports [26], the yield point (which is identified by the point of inflexion between the origin and the maximum stress point on the stress–strain curve) was used to determine the yield strength (σ_Y), yield strain (ε_Y), stiffness (E), and strain energy density to yielding (u_Y), otherwise known as resilience energy). The maximum stress point was identified to correspond to the fracture strength (σ_U) and the fracture strain (ε_U, otherwise known as extensibility); the strain energy density to fracture (u_F, otherwise known as fracture toughness) was determined up to this point [26,43].

2.3. Statistical Analysis

Two-factor ANOVA was primarily used to test our hypotheses (Section 1), i.e., that the particle type and concentration have significant effects on the respective mechanical properties and that interactions occur between the two factors, at an alpha level of 0.05. When interaction effects caused a masking of the results (such as in the case of CS/HA versus CS/POSS), a Student's t-test was used to analyse for differences at the respective levels. A test was regarded as significant when p value < 0.05.

2.4. Weibull Model

To analyse the mechanical reliability of the composite fibres, the Weibull model was applied to evaluate the mechanical data. Let β and σ_0 represent the Weibull modulus and the characteristic strength, respectively. Adapting the method for fracture analysis (probabilistic approach) from a previous study on CS/HA with respect to particle concentration versus crystallisation temperature [34], according to Weibull's empirical law [44], the cumulative distribution function (C) of the yield/fracture stress of the CS composite fibre, σ, for determining failure due to flaws is given by $C(\sigma) = 1 - \exp(-[\sigma/\sigma_0]^\beta)$.

To apply the Weibull law to the probabilistic analysis of the yield/fracture of the CS composite fibre, first one notes that β quantifies the variability of σ; low β values correspond to high variability and vice versa [44]. Second, one notes that σ_0 is the stress value at which 63% of the fibres have yielded/fractured [44]. Normally, for the convenience of analysis the C function is replaced by the reliability function, $R (= 1 - C)$, which describes the proportion of the population of specimens sampled that survive at fracture stress σ. Thus the expression for R is given by

$$R(\sigma) = \exp(-\{\sigma/\sigma_0\}^\beta). \tag{1}$$

Additionally, the median rank position,

$$M = \{i - 0.3\}\{n + 0.4\} - 1, \tag{2}$$

where n represents the size of the treatment group and i is the position of the corresponding σ, is numerically identified with the R. It then follows that M is used to compute the R as an intermediate step in the Weibull analysis.

We have adopted the following practical approach to evaluate β and σ_0 for the different treatment groups, namely CS/HA versus CS/POSS and particle concentration levels. First, we determined the M for each experimentally derived value of σ (i.e., σ_Y or σ_U). This was carried out by ranking the magnitudes of the σ (i.e., σ_Y or σ_U) in ascending magnitude. The corresponding estimates of M were evaluated using Equation (2). Second, we fitted straight lines to the so-called Weibull plot of $\log(\log(1/\{1 - M(\sigma)\}))$ versus $\log(\sigma)$ for each group. Finally, the value of β was identified with the slope of the respective straight lines; the value of σ_0 was found after equating $-\beta \log(\sigma_0)$ to the y-intercept of the straight line.

3. Results

3.1. Fracture Morphology

The overall fracture morphologies for the CS/HA and CS/POSS fibres (Figure 1) appear somewhat similar. The fractured planes are near-perpendicular to the axis of the fibre, suggesting that the fibres fail by brittle fracture. All fibres exhibit similar fissures and projection features. These could be attributed to a combination of the following modes of failure: (A) detachment (i.e., pullout) of the HA or POSS particles from the chitosan matrix (inter-granular failure); (B) fracture of the HA or POSS particles (trans-granular failure). POSS compounds are in the form of nanoparticles; in principle these could be observable in the fibres using a SEM with high magnification. So far, these are the best images that one could derive from SEM imaging. Here we also noted that the images reported elsewhere by other researchers reveals POSS in large particulates on the order of 1 μm, when blended in a chitin matrix [8]. Note that the ribbon-like profile of the fibre cross section was the result of lateral compressive forces that were generated by the extruded fibre by winding around a bobbin during the wet spinning process. Of note, the ribbon-like profile could present an advantage as it possesses a greater surface area for cell adhesion. Additionally, if the fibres could be laid down in the form of a mesh, the 'pores', i.e., the space between the fibres, need to be sufficiently large for cells to migrate into the mesh, where they eventually become bound to the surfaces in the scaffold. Nevertheless, the design of a scaffold mesh made from such CS fibres is promising but further work is required to optimise the fibre surface area per unit volume (A_{SA}) and pore size (D_{pore}) [45,46].

We have not attempted to mesh our fibres as the main focus is on the properties of the single fibre. In our samples, the ribbon-like profile provides a broad surface of about 2pi × 40/2 (μm) = 126 μm. To this end, we could expect that the pore sizes are on the order of 100 μm or more. Fibroblasts are on the order of magnitude of 40–60 μm [47]. Osteoblasts and chondrocytes are on the order of 10 μm [48]. More importantly, the sizes of these cells are consistent with the dimensions of the fibre thickness, as well as the predicted size of the pore.

If a scaffold (meshed) is to be fabricated from CS/HA or CS/POSS fibres, the first issue to address is how to lay down the fibres with regards to having a fibre orientation predominated by primary fibres in one direction (to provide certainty in strength and stiffness) and randomly oriented secondary fibres. What then is the mechanical property of this configuration? It turns out that for such a design, the strength of the mesh (which is related to the probability of rupturing of a fibre section) is dictated by the Weibull distribution of stresses (Section 3.4). The overall tensile strength of the mesh (when gripped at both extreme ends to stretch to rupture) would be lower than that of a single fibre [49].

3.2. Effects of Particle Type and Concentration on the Elastic Properties

To begin, we have found significant differences ($p < 0.05$) for the respective E (Figure 2A) and σ_Y (Figure 2B) versus particle concentration and particle type. No significant interaction between the two factors is observed ($p > 0.05$). For the main effects of particle type, the mean E and σ_Y of the CS/POSS fibres are higher than those of the CS/HA fibres. As for the main effects of particle concentration, the mean E and σ_Y are highest at 3% (w/w) and lowest at 9% (w/w) particle concentration. It can be concluded that (A) the CS/POSS fibres are stiffer and have higher yield strength than those of CS/HA; and (B) an optimal particle concentration that leads to the highest stiffness and yield strength occurs at 3% (w/w).

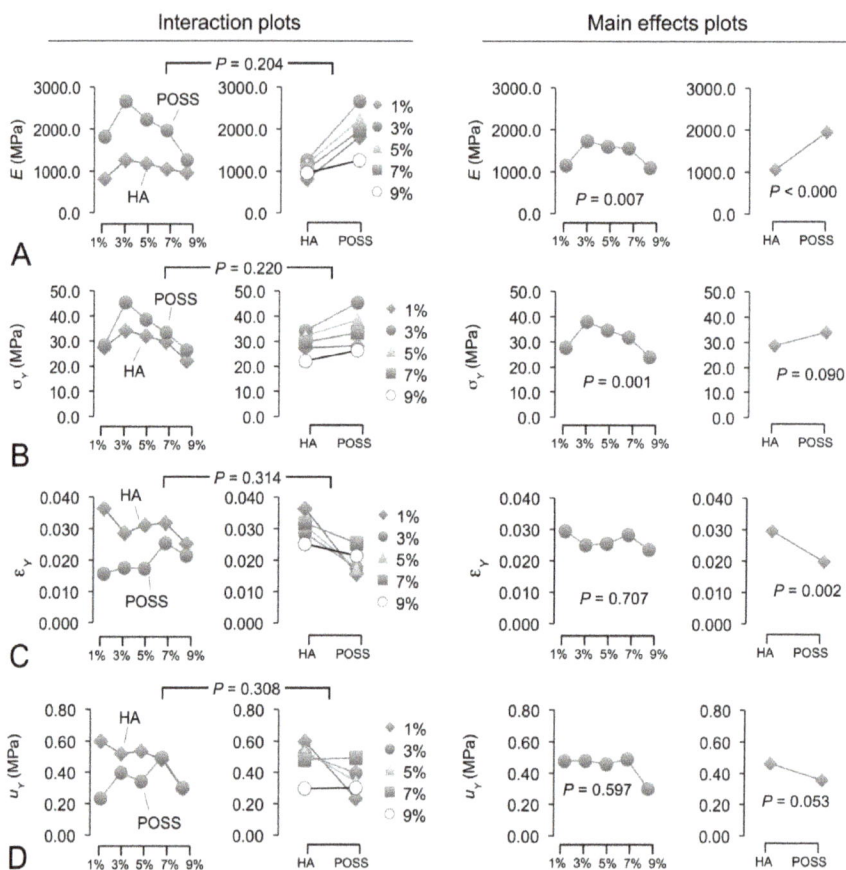

Figure 2. Interaction (**left panel**) and main effects (**right panel**) of particle concentration (% w/w) and particle type on the elasticity-related properties of chitosan-based fibres reinforced by hydroxyapatite (HA) versus polyhedral oligomeric silsesquioxane (POSS) particles. Symbols: (**A**) E, elastic modulus; (**B**) σ_Y, yield stress; (**C**) ε_Y, yield strain; (**D**) u_Y, strain energy density for resilience.

The p value for the effects of particle type on ε_Y is small ($p = 0.002$), revealing strong evidence of the influence of particle type on ε_Y (Figure 2C). However, this is not the case for ε_Y versus particle concentration ($p = 0.707$). Since the p value for the interaction between particle concentration and particle type is much greater than 0.05 (interaction $p = 0.314$), this suggests that there is no evidence of an interaction between the factors. We conclude that ε_Y is sensitive to variation in particle type but not to particle concentration. In considering the main effects of particle type, it is observed that the mean ε_Y from the CS/HA fibre is higher than that of the CS/POSS fibre. We conclude that CS/HA fibres are more deformable, i.e., they yield at larger strains than CS/POSS fibres.

As for u_Y, we note that the p values for particle type ($p = 0.053$) and particle concentration ($p = 0.597$) are both greater than 0.05 (although the former may be regarded as marginal). Thus, there is no evidence that u_Y is sensitive to particle type and concentration (Figure 2D). The p value for the interaction is much greater than 0.05 ($p = 0.308$), showing that there is no evidence of an interaction between the factors. Altogether, this suggests that u_Y is not sensitive to variations in particle concentration and type.

3.3. Effects of Particle Type and Concentration on Fracture Properties

With regards to σ_U, the ANOVA results (Figure 3B) reveal a significant interaction between particle type and particle concentration ($p < 0.001$). Thus the main effects may not be interpreted independently of one another. Figure 3A shows that σ_U is sensitive to variations in particle concentration ($p < 0.001$) but not particle type ($p = 0.216$). It appears that the optimal particle concentration occurs at 3% or 7% (w/w) depending on the particle type. The main effects due to particle type could be masked by interaction effects and warrant further analysis. To address the effects arising from interaction, a two-sample t-test was conducted to investigate the differences in σ_U between the CS/HA and CS/POSS at the respective particle concentration levels. The results of the t-test reveal no significant difference in σ_U between the CS/HA and CS/POSS for all levels of particle concentration except at 7% (w/w) ($p < 0.001$)—the mean σ_U of the CS/HA fibres is smaller than that of CS/POSS fibres at 7% (w/w).

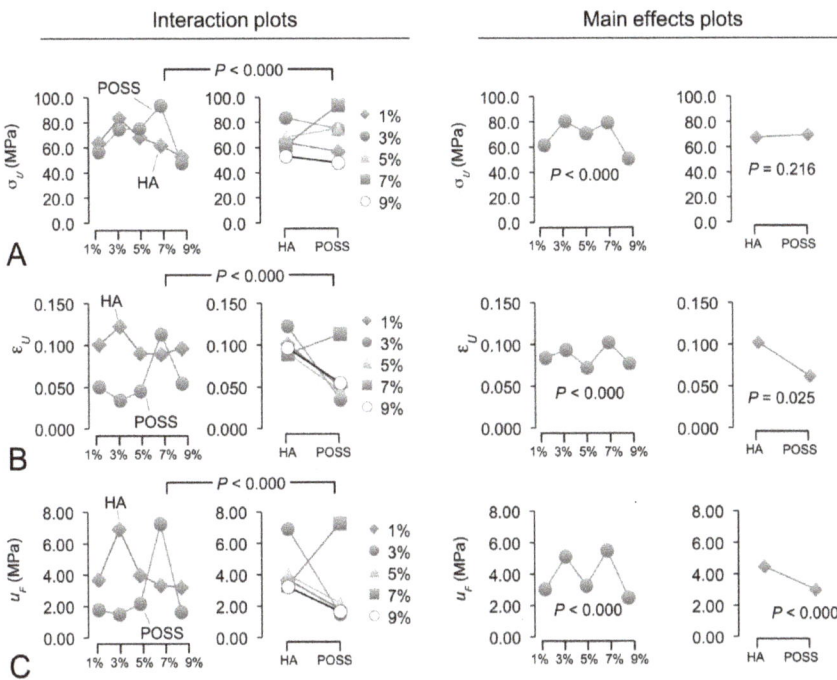

Figure 3. Interaction plots (**left panel**) and main effects (**right panel**) plots of particle concentration (% w/w) and particle type on the fracture-related properties of chitosan-based fibres reinforced by hydroxyapatite (HA) versus polyhedral oligomeric silsesquioxane (POSS) nanoparticles. Symbols: (**A**) σ_U, fracture strength; (**B**) ε_U, extensibility; (**C**) u_F, strain energy density to fracture.

With regards to ε_U, a significant interaction between particle concentration and type occurs ($p < 0.001$). Thus the main effects may not be interpreted independently of one another. Here, significant differences are observed for ε_U versus particle type ($p < 0.001$) and particle concentration ($p = 0.025$) (Figure 3B)—the ε_U is sensitive to variations in particle type and particle concentration but the effects of particle type on ε_U is modified by particle concentration and vice versa. We conclude that (1) the mean ε_U from the CS/HA fibre is higher than that of the CS/POSS fibre (except at 7% w/w, where both result in similar magnitudes of ε_U); and (2) the optimal particle concentration resulting in the highest ε_U occurs at 7% (w/w).

Finally, with regards to u_F, the p value for the interaction between particle type and concentration is very small ($p < 0.001$), showing that there is evidence of an interaction between the two factors. Thus, the main effects of particle type and particle concentration on u_F may not be interpreted independently of one another. Nevertheless, significant differences are observed for u_F versus particle type ($p = 0.001$) and particle concentration ($p < 0.001$) (Figure 3C). In considering the main effects of particle type, it is observed that the mean u_F from the CS/HA fibres is about 40% higher than that of the CS/POSS fibres. For the consideration of the main effects of particle concentration, the mean u_F appear to peak at 3 or 7% (w/w), depending on the particle type. We conclude that the optimal particle concentration, corresponding to maximum u_F occurs at 3 and 7% (w/w) for HA- and CS/POSS fibres, respectively.

3.4. Mechanical Reliability

Table 1 lists the values of the Weibull modulus, β, and the characteristic strength, σ_0, of the CS composite fibre reinforced by HA and POSS nanoparticles, associated with the yield strength and fracture strength parameters. Inspection of the values of β in the table reveals the following trends. With regards to the yielding of CS/HA composite fibre, it is observed that β increases with the increase in POSS concentration, peaks at 7% w/w, and decreases somewhat thereafter. As for the yielding of the CS/POSS composite fibre, it is observed that β increases rapidly with an increase in POSS and peaks at 3% w/w (as compared to 7% w/w for the case of CS/HA), followed by a decrease in β, with increasing POSS concentration. On the other hand, with regards to the fracture of the CS fibre, the β from both CS/HA and CS/POSS appears to fluctuate with increasing particle concentration, and no appreciable trend is observed.

Table 1. Analysis of the Weibull modulus, β, and the characteristic strength, σ_0, of chitosan (CS) fibres reinforced by hydroxyapatite (HA) and polyhedral oligomeric silsesquioxane (POSS) nanoparticles.

Particle Concentration % w/w	CS/HA				CS/POSS			
	Yield Strength		Fracture Strength		Yield Strength		Fracture Strength	
	β (MPa)	σ_0 (MPa)	β (MPa)	σ_0 (MPa)	β (MPa)	σ_0 (MPa)	β (MPa)	σ_0 (MPa)
1	4.58	37.53	7.21	69.43	2.40	31.76	13.00	58.96
3	4.80	38.31	12.90	86.26	7.90	48.11	4.70	85.35
5	4.90	35.57	9.80	70.46	4.40	42.52	3.50	83.81
7	5.30	32.80	15.80	62.76	2.60	37.17	13.30	97.41
9	3.70	24.27	7.80	55.30	2.50	32.46	9.40	50.68

Inspection of the values of σ_0 in the table reveals the following trends. With regards to the yielding of CS composite fibre, both CS/HA and CS/POSS show that σ_0 increases with an increase in the particle concentration, peaks at 3% w/w, and decreases somewhat thereafter. On the other hand, as for the fracture of the CS/HA fibre, σ_0 increases with increasing HA concentration, peaks at 3% w/w, and decreases thereafter. In contrast, for the CS/POSS fibre, σ_0 increases with increasing POSS concentration, peaks at 7% w/w, and decreases thereafter.

Figure 4 shows plots of R versus σ for the CS/HA and CS/POSS fibres corresponding to the cases of yielding and fracture. These plots illustrate the sensitivity of the reliability to the β and σ_0 parameters at the respective levels of particle concentration and type. With the exceptions of the yielding of CS/POSS at 1%, 7% and 9% w/w and the fracture of CS/POSS fibre at 3% and 5% w/w, the form (i.e., parameterized by β) of the curves corresponding to CS/HA and CS/POSS features a very narrow spread of strength variability, for the respective case of yielding and fracture. The scatter of the curves deserves some attention. With regards to the respective HA and POSS particles, over the range of particle concentration considered here, the curves for the yielded CS fibre (Figure 4A (C)) are less scattered compared to those of fractured CS fibres (Figure 4B (D)). The implications of these results are discussed in Section 4.2.

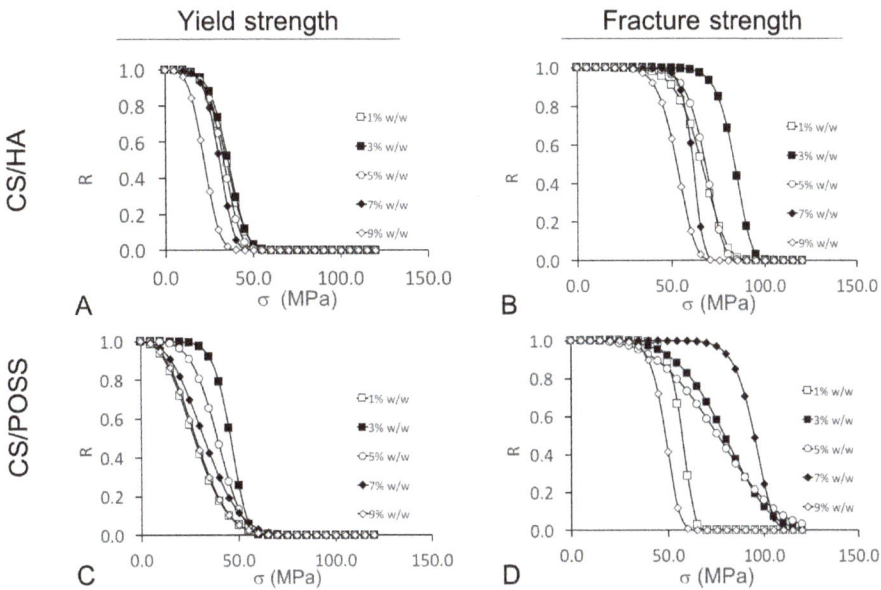

Figure 4. Plot of reliability function, R (a dimensionless quantity), versus stress (σ, MPa) in the chitosan-based composite fibre, reinforced by hydroxyapatite (HA), for the case of (**A**) yield strength and (**B**) fracture strength, and polyhedral oligomeric silsesquioxane (POSS), for the case of (**C**) yield strength and (**D**) fracture strength, in the presence of varying particle concentrations (% w/w). Note: σ_0 and β represent the characteristic strength and the Weibull modulus, respectively.

4. Discussion

4.1. Study Findings

In this study, both CS/POSS and CS/HA fibres were synthesized by a wet spinning method, with varying particle concentrations from 1% to 9% (w/w). Wet spinning is a simple and effective method that involves a combination of rheological and diffusional mechanisms to synthesize fibres with controlled thickness [25,26,50,51]. Methods such as melt spinning are not feasible as the CS polymers degrade upon heating [25,51]. It is important to note that the wet spinning method can produce useful fibres because the flow action during extrusion helps to align both the CS polymer molecules and the nanoparticles in the direction of the fibre axis, thus contributing to axial reinforcement [26]. From the perspective of tissue engineering scaffolds, it makes sense to synthesize CS fibres of micrometre thickness for several reasons, namely they can lead to large surface area for cell attachment as well as enhanced interconnected pore architecture that provides pathways for the diffusion of gases, transportation of nutrients, and migration of cells [11,52–54].

According to some reports, isolated particles of HA and POSS are physically distinguished by their shapes: HA are generally described as having a needle-like (generally rods or with straight-taper ends) profile [55,56] while POSS particles may be described as having globular profiles [57]. While these descriptions are not based on a detailed analysis of the particle shapes, the precise form of these particles (at least for HA particles [25]) is expected to depend on the precipitation temperature, which ranges from ambient [58,59] to 40 °C [29,50,51,60]. More importantly, both particle types provide a unique perspective to study the effects of non-uniform cylindrical profiles of a nanoparticle-reinforced composite and its mechanical properties [30]. Although the exact forms of these particles in the composite are not observable, predictions from finite element models of composites reinforced by ellipsoidal particles and needle-like particles with straight tapered ends reveal that the stress uptake in these particles is somewhat evenly distributed throughout during elastic loading [61]. On the other hand, uniform cylindrical (i.e., rod shape) particle-reinforced composites feature a peak stress at the particle centre; the stress decreases non-linearly to zero at the particle ends [62]. In principle, the non-uniform cylindrical particles are less likely to break owing to lower stress concentrations, and hence are more effective than uniform cylindrical ones for composite reinforcement [62,63] Nonetheless, these predictions have yet to be supported by experimental results.

There are some reports of 'hybrid' composites, notably from the study based on computer modelling of a composite comprising (i.e., non-fibre-like) particles and fibres blended in polymer composites [64], and banana/sisal fibres blended in epoxy resin [65]. Of note, the mechanical properties of a banana/sisal fibre-reinforced epoxy resin composite have been evaluated based on the rule-of-mixture for hybrid composites [49]. However, to the best of our knowledge, there is no study hybridizing chitosan by blending with POSS and HA particles. This would be an interesting study with regards to optimising the chitosan composite mechanical properties.

In principle, the magnitudes of the mechanical properties (such as strength and stiffness) are expected to increase linearly with particle concentration in good order-of-magnitude agreement with the simple rule-of-mixture for strength and stiffness [30]. In practice, composites reinforced by nanoparticles (such as those shown in this study for the CS/HA and CS/POSS, and in other studies, namely (A) SiO_2 particles reinforcing polyimide [1], (B) $BaSO_4$ particles reinforcing polypropylene [66], and (C) carbon nanotubes reinforcing ceramic [67]) do not follow the rule-of-mixture at large particle concentration levels. In fact, there exists a particle concentration level beyond which a further increase in particle concentration leads to a diminution in the mechanical properties. This effect is often attributed to the extensive agglomeration of nanoparticles beyond the optimal particle concentration [1,30,68]. Further discussion of the underlying causes of agglomeration is found in Section 4.2. Nevertheless, the low particle concentration is a cause for concern as it severely constrains the tunability of the mechanical properties of the composite to a narrow range of particle concentrations. Interestingly, the peak value of different mechanical parameters appears to differ. For instance, the highest stiffness occurs at 3% w/w but the highest fracture strength occurs at 7% w/w. This means that if the material designer is required to design the CS composite fibre for high fracture strength, the trade-off would be lower stiffness.

An interesting study has been carried out by Aranaz et al. [69] to investigate the type of calcium phosphate formed in the CSCaP monoliths. There are two key findings according to this study [69]. (A) Calcined samples showed a pattern similar to enamel (highly crystalline apatite mineral, ca. 96 wt. %), and non-calcined samples showed a pattern similar to dentin (partially amorphous apatite, which also contains an organic matrix and water). (B) The amount of calcium phosphate in the CSCaP composites can significantly influence the efficacy of cell proliferation: it is found that higher calcium phosphate content in the composite enhances cell attachment and proliferation and vice versa. With regards to our CS fibres, it would be interesting as part of future work to assess the extent of the crystallinity in the CS fibres. This could come together with plans for cell studies. With regards to cell studies, one key area of interest is in developing a consistent technique to lay down the fibres into a mesh. The study of osteoblast and chondrocyte cells in tissue engineering with regards to cell adhesion, spreading, and viability is an ongoing interest [48]. With regards to our CS fibres, one future work of interest would be to investigate how these cells respond to the fibrous mesh when they are seeded into the mesh. A method will be developed to enable consistency in the seeding process using an in-house developed automated cell dispensing machine; such a machine may be modified from an open-sourced 3D printer [70].

Although the mechanical properties of the CS/HA composite are well established and a few reports have been published on the CS/POSS composite, most reports have been concerned with singly applied treatments. In other words, these experiments have attempted to investigate only one synthesis parameter, notably the effects of particle concentration [25,26]. Here, the results from the two-factor ANOVA reveal that particle concentration and particle type do interact. With regards to the sensitivity of the mechanical properties of CS composite fibre to particle concentration and particle type, the findings are listed as follows: (A) Only the fracture-related mechanical parameters, i.e., σ_U, ε_U and u_F, are sensitive to interaction effects. Since these parameters are also sensitive to particle concentration and type, the trending variation of the values of each of these parameters with varying particle concentration (type) depends on particle type (concentration). As for the elasticity-related parameters (i.e., E, σ_Y, ε_Y and u_Y), the results suggest that the main effects (where applicable) of particle concentration and particle type on these parameters may be interpreted independently of one another. (B) With regards to effects from particle type, by and large the σ_Y and E of CS/POSS fibres are higher than CS/HA fibres but CS/HA fibres possess higher ε_Y, ε_U and u_F than CS/POSS fibres. This suggests that CS/POSS fibres are stronger and stiffer than CS/HA fibres but CS/HA fibres are not only less brittle but also tougher than CS/POSS fibres. The u_Y is not sensitive to particle type. (C) With regards to effects from particle concentration, only E and σ_Y reveal a clear trending increase up to an optimal particle concentration; thereafter a trending decrease occurs. The optimal particle concentration differs for different mechanical properties. In particular, E and σ_Y peak at 3% (w/w). The ε_Y and u_Y are not sensitive to particle concentration. Interestingly, due to the presence of interaction between the particle concentration and particle type, σ_U, ε_U and u_F peak at 3% or 7% (w/w), depending on particle type.

The E, u_Y and u_F may be regarded as 'derived parameters', defined in terms of both stress and strain components. By considering the results from the main effect study (Sections 3.2 and 3.3), we infer that (A) the stress component predominates in E as demonstrated by σ_Y; (B) the strain component (from initial loading until ε_U) predominates in u_F, as demonstrated by ε_Y. While further discussion concerning the basis for the absence of sensitivity of the u_Y of chitosan composite fibre to nanoparticles of HA and POSS could be valuable, the answer is not as clear-cut as we might wish. Nevertheless, the underpinning arguments suggest that the mechanics of nanoparticles of HA for modulating the stress and strain (by directing low stress uptake at high strains) is in direct contrast to that of POSS (by directing high stress uptake at low strains). However, in either case, these lead to similar results arising from a similar amount of strain energy absorbed per unit volume (i.e., u_Y) causing yielding in the fibre.

4.2. Design for Reliability

It is possible that defects in the CS fibre may originate from agglomerates of HA or POSS, as well as localised non-uniform dispersion of the respective particles within the CS fibre—these defects then act as stress intensifiers. Failure in the fibre may begin at these stress intensifier site as the load on the fibre increases; yielding occurs when chemical bonds within the defects are partially disrupted but rupture, i.e., formation of new crack surfaces, occurs when all the chemical bonds at a particular site are dissociated.

The reliability predictions reveal that the CS/HA composite fibres have more consistent (in other words, narrower spread of σ values) yield and fracture strengths as compared to CS/POSS composite fibres. Thus this could be attributed to two factors: (A) the agglomeration of particles and (B) the directionality of the particles. The agglomerates could compromise the mechanical properties of the composite fibre. As these agglomerates are weakly bonded together, the dissociation of these agglomerates could in turn contribute to increased unpredictability in the yield and fracture strength of the CS/POSS fibre. Additionally, as the POSS particles are described as having globular profiles (Section 4.1), this means that in some POSS particles the long axis may be comparable in length to the short axis. Thus, during the processing of the CS/POSS fibre, as the POSS particles flow though the tube and spinneret, the flow effect may not be sufficient to cause the long axis of all POSS particles to point in the direction of the fibre axis. Consequently, the directionality of the POSS particles is less defined than that of HA particles, which are described as having needle-like profiles. The directionality of the particulates plays an important role in determining the mechanical properties of the CS composite fibre. According to the principles of particle-reinforced composites, a composite that contains well-aligned particles will have higher strength when an external load is applied in the direction of these particles [26]. For composites that contain particles that are not well aligned, only a proportion of the particles that are aligned in the direction of the applied load would be able to provide effective reinforcement by taking up stress from the matrix [26].

The possible contribution of particle agglomeration to the decrease in the magnitude of the respective mechanical parameters, at high particle concentration levels, is an issue of great concern [68]. As these agglomerates are weakly bonded, individual agglomerates dissociate easily into smaller particles under high load—smaller particles, with shorter length, may be less effective at taking up high stress. Exactly how the weak bonds affect both the elasticity- and fracture-related mechanical properties of the chitosan composite is not clearly understood and is a subject for further study.

As noted in Section 4.1, in cases where the interaction between particle type and concentration is significant, the optimal particle concentration for each mechanical parameter differs depending on the particle type. This appears to reflect the varying degree of agglomeration by each particle type and may have to do with the degree of the proximity of particles within an agglomerate. Accordingly, from computational studies on particle–particle proximity [68], it is inferred that (A) the interphase material surrounding a particle could overlap with the interphase material from neighbouring particles, and (B) the mechanical properties of the nanoparticulate-reinforced composite could depend on the degree of the interphase-interphase overlap. It then follows that particle clustering increases appreciably at high particle concentration levels. Consequently, this leads to decreases in the E of the nanoparticulate-reinforced composites [68].

The non-linear relationship between the mechanical property of a composite and the particle concentration, and the existence of an optimal particle concentration that results in the characteristic peak value of the mechanical parameter, means that predicting the mechanical properties of these composites may not be as straightforward as we would like. As pointed out in Section 4.1, these non-linear effects are found in many particle-reinforced composite systems. For instance, polypropylene reinforced by $BaSO_4$ nanoparticles reveals an increasing σ_Y with increasing particle concentration up to 10% (w/w); thereafter a trending decrease occurs with increasing particle concentration [66]. Polyimide reinforced by SiO_2 particles reveals an increasing σ_U with particle concentration for up to 5% (w/w); thereafter σ_U decreases with increasing particle concentration [1]. Epoxy reinforced by hydrated $Al(OH)_3$ nanoparticles results in increasing G (fracture toughness in kJ/m^2) with particle concentration from 20% to 30% (w/w); thereafter a decreasing G with particle concentration was observed [71]. Nevertheless, there are exceptions, e.g., dentin composites reinforced by hydroxyapatite nanoparticles reveal a monotonically increasing (flexural) E with increase in particle concentration (the maximum particle concentration studied was 15% w/w [72]; it remains to be seen if E also peaks at a certain particle concentration thereafter before decreasing. Recently, simple constitutive equations based on phenomenological arguments have been proposed to model these non-linear effects for fracture strength and stiffness [49]. The underlying basis of the model suggests that the initial linear increase in the magnitude of the respective mechanical property with increase in particle concentration follows the rule-of-mixture [49]. Beyond a certain particle concentration, the matrix mechanical property dominates so that the mechanical property experiences decrease in magnitude with increasing particle concentration [49]. Nevertheless, this model has to be investigated further before it can be used in practical applications.

5. Conclusions

It is shown in this study that only the fracture-related properties were sensitive to effects arising from the interaction between the particle type and concentration. Most mechanical properties are sensitive to particle type; for instance, the stiffness and yield strength of CS/POSS fibres are higher than those of CS/HA fibres. On the other hand, the yield strain, extensibility and fracture toughness of CS/POSS fibres are lower than those of CS/HA fibres. Most mechanical properties are sensitive to varying particle concentrations. For some mechanical properties, namely stiffness and yield strength, these feature a trending increase to a peak value (the optimal particle concentration) and a trending

J. Compos. Sci. **2017**, *1*, 9

decrease thereafter, with increasing particle concentration. For the others, namely fracture strength, strain at fracture and fracture toughness, the sensitivity is significant but no trending increase/decrease is sustained over the particle concentration range investigated here. Predictions of the reliability of the CS composite fibres reveals that CS/HA composite fibres have a more consistent (in other words, narrower spread of σ values) yield and fracture strength as compared to CS/POSS composite fibres.

Acknowledgments: The authors gratefully acknowledge the help of Shir Lin Chew with the experimental part of this work. The authors also gratefully acknowledge the initial discussions with San Hein on the setup of the wet spinning system and the synthesis process of a chitosan composite fibre.

Author Contributions: K.W. and K.L.G. conceived and designed the experiments. K.W. and K.L.G. performed the experiments; K.W., P.P., R.T.D.S. and K.L.G. analyzed the data; K.W. contributed reagents/materials/analysis tools; K.W., P.P., R.T.D.S. and K.L.G. wrote the paper.

Conflicts of Interest: The authors declare no conflict of interests

References

1. Fu, S.Y.; Feng, X.Q.; Lauke, B.; Mai, Y.W. Effects of particle size, particle/matrix interface adhesion and particle loading on mechanical properties of particulate-polymer composites. *Compos. Part B* **2008**, *39*, 933–961. [CrossRef]
2. De Silva, R.T.; Pasbakhsh, P.; Goh, K.L.; Mishnaevsky, L. 3-D computational model of poly (lactic acid)/halloysite nanocomposites: Predicting elastic properties and stress analysis. *Polymer* **2014**, *55*. [CrossRef]
3. De Silva, R.T.; Soheilmoghaddam, M.; Goh, K.L.; Wahit, M.U.; Bee, S.A.H.; Chai, S.-P.; Pasbakhsh, P. Influence of the processing methods on the properties of poly (lactic acid)/halloysite nanocomposites. *Polym. Compos.* **2016**, *37*, 861–869. [CrossRef]
4. Khor, E.; Lim, L.Y. Implantable applications of chitin and chitosan. *Biomaterials* **2003**, *24*, 2339–2349. [CrossRef]
5. Khalil, H.P.S.A.; Saurabh, C.K.; Adnan, A.S.; Fazita, M.R.N.; Syakir, M.I.; Davoudpour, Y.; Rafatullah, M.; Abdullah, C.K.; Haafiz, M.K.M.; Dungani, R. A review on chitosan-cellulose blends and nanocellulose reinforced chitosan biocomposites: Properties and their applications. *Carbohydr. Polym.* **2016**, *150*, 216–226. [CrossRef]
6. Wysokowski, M.; Bazhenov, V.V.; Tsurkan, M.V.; Galli, R.; Stelling, A.L.; Stöcker, H.; Kaiser, S.; Niederschlag, E.; Gärtner, G.; Behm, T.; et al. Isolation and identification of chitin in three-dimensional skeleton of Aplysina fistularis marine sponge. *Int. J. Biol. Macromol.* **2013**, *62*, 94–100. [CrossRef] [PubMed]
7. Wysokowski, M.; Petrenko, I.; Stelling, A.L.; Stawski, D.; Jesionowski, T.; Ehrlich, H. Poriferan chitin as a versatile template for extreme biomimetics. *Polymers* **2015**, *7*, 235–265. [CrossRef]
8. Wysokowski, M.; Materna, K.; Walter, J.; Petrenko, I.; Stelling, A.L.; Bazhenov, V.V.; Klapiszewski, Ł.; Szatkowski, T.; Lewandowska, O.; Stawski, D.; et al. Solvothermal synthesis of hydrophobic chitin-polyhedral oligomeric silsesquioxane (POSS) nanocomposites. *Int. J. Biol. Macromol.* **2015**, *78*, 224–229. [CrossRef] [PubMed]
9. Obara, K.; Ishihara, M.; Ishizuka, T.; Fujita, M. Photocrosslinkable chitosan hydrogel containing fibroblast growth factor-2 stimulates wound healing in healing-impaired db/db mice. *Biomaterials* **2003**, *24*, 3437–3444. [CrossRef]
10. Cheng, M.; Deng, J.; Yang, F.; Gong, Y.; Zhao, N.; Zhang, X. Study on physical properties and nerve cell affinity of composite films from chitosan and gelatin solutions. *Biomaterials* **2003**, *24*, 2871–2880. [CrossRef]
11. Malheiro, V.N.; Caridade, S.G.; Alves, N.M.; Mano, J.F. New poly (e-caprolactone)/chitosan blend fibers for tissue engineering applications. *Acta Biomater.* **2010**, *6*, 418–428. [CrossRef] [PubMed]
12. Alves, N.M.; Mano, J.F. Chitosan derivatives obtained by chemical modifications for biomedical and environmental applications. *Int. J. Biol. Macromol.* **2008**, *43*, 401–414. [CrossRef] [PubMed]
13. Di Martino, A.; Sittinger, M.; Risbud, M.V. Chitosan: A versatile biopolymer for orthopaedic tissue-engineering. *Biomaterials* **2005**, *26*, 5983–5990. [CrossRef] [PubMed]

14. Albanna, M.Z.; Bou-Akl, T.H.; Blowytsky, O.; Walters, H.L.; Matthew, H.W.T. Chitosan fibers with improved biological and mechanical properties for tissue engineering applications. *J. Mech. Behav. Biomed. Mater.* **2013**, *20*, 217–226. [CrossRef] [PubMed]

15. Liu, M.; Zhang, Y.; Wu, C.; Xiong, S.; Zhou, C. Chitosan/halloysite nanotubes bionanocomposites: Structure, mechanical properties and biocompatibility. *Int. J. Biol. Macromol.* **2012**, *51*, 566–575. [CrossRef] [PubMed]

16. De Silva, R.T.; Pasbakhsh, P.; Goh, K.L.; Chai, S.-P.; Ismail, H. Physico-chemical characterisation of chitosan/halloysite composite membranes. *Polym. Test.* **2013**, *32*, 265–271. [CrossRef]

17. Govindasamy, K.; Fernandopulle, C.; Pasbakhsh, P.; Goh, K.L. Synthesis and characterisation of electrospun chitosan membranes reinforced by halloysite nanotubes. *J. Mech. Med. Biol.* **2014**, *14*, 1450058. [CrossRef]

18. Goh, K.L.; Listrat, A.; Béchet, D. Hierarchical mechanics of connective tissues: Integrating insights from nano to macroscopic studies. *J. Biomed. Nanotechnol.* **2014**, *10*, 2464–2507. [CrossRef]

19. Wong, W.L.E.; Joyce, T.J.; Goh, K.L. Resolving the viscoelasticity and anisotropy dependence of the mechanical properties of skin from a porcine model. *Biomech. Model. Mechanobiol.* **2016**, *15*, 433–446. [CrossRef] [PubMed]

20. Goh, K.L.; Holmes, D.F.; Lu, H.Y.; Richardson, S.; Kadler, K.E.; Purslow, P.P.; Wess, T.J. Ageing changes in the tensile properties of tendons: Influence of collagen fibril volume. *J. Biomech. Eng.* **2008**, *130*, 21011. [CrossRef] [PubMed]

21. Goh, K.L.; Chen, Y.; Chou, S.M.; Listrat, A.; Bechet, D.; Wess, T.J.J. Effects of frozen storage temperature on the elasticity of tendons from a small murine model. *Animal* **2010**, *4*, 1613–1617. [CrossRef] [PubMed]

22. Goh, K.L.; Holmes, D.F.; Lu, Y.; Purslow, P.P.; Kadler, K.E.; Bechet, D.; Wess, T.J. Bimodal collagen fibril diameter distributions direct age-related variations in tendon resilience and resistance to rupture. *J. Appl. Physiol.* **2012**, *113*, 878–888. [CrossRef] [PubMed]

23. Yeo, Y.L.; Goh, K.L.; Kin, L.; Wang, H.J.; Listrat, A.; Bechet, D. Structure-property relationship of burn collagen reinforcing musculo-skeletal tissues. *Key Eng. Mater.* **2011**, *478*, 87–92. [CrossRef]

24. Goh, K.L.; Chen, S.Y.; Liao, K. A thermomechanical framework for reconciling the effects of ultraviolet radiation exposure time and wavelength on connective tissue elasticity. *Biomech. Model. Mechanobiol.* **2014**. [CrossRef] [PubMed]

25. Xie, J.Z.; Hein, S.; Wang, K.; Liao, K.; Goh, K.L. Influence of hydroxyapatite crystallization temperature and concentration on stress transfer in wet-spun nanohydroxyapatite-chitosan composite fibres. *Biomed. Mater.* **2008**, *3*, 2–6. [CrossRef] [PubMed]

26. Chew, S.L.; Wang, K.; Chai, S.P.; Goh, K.L. Elasticity, thermal stability and bioactivity of polyhedral oligomeric silsesquioxanes reinforced chitosan-based microfibres. *J. Mater. Sci. Mater. Med.* **2011**, *22*, 1365–1374. [CrossRef] [PubMed]

27. Blanco, I.; Abate, L.; Bottino, F.A. Synthesis and thermal properties of new dumbbell-shaped isobutyl-substituted POSSs linked by aliphatic bridges. *J. Therm. Anal. Calorim.* **2014**, *116*, 5–13. [CrossRef]

28. Guo, Y.-P.; Guan, J.-J.; Yang, J.; Wang, Y.; Zhang, C.-Q.; Ke, Q.-F. Hybrid nanostructured hydroxyapatite–chitosan composite scaffold: Bioinspired fabrication, mechanical properties and biological properties. *J. Mater. Chem. B* **2015**, *3*, 4679–4689. [CrossRef]

29. Yamaguchi, I.; Tokuchi, K.; Fukuzaki, H.; Koyama, Y.; Takakuda, K.; Monma, H.; Tanaka, J. Preparation and microstructure analysis of chitosan/hydroxyapatite nanocomposites. *J. Biomed. Mater. Res.* **2001**, *55*, 20–27. [CrossRef]

30. De Silva, R.; Pasbakhsh, P.; Qureshi, A.J.; Gibson, A.G.; Goh, K.L. Stress transfer and fracture in nanostructured particulate-reinforced chitosan biopolymer composites: Influence of interfacial shear stress and particle slenderness. *Compos. Interfaces* **2014**, *21*, 807–818. [CrossRef]

31. Peniche, C.; Solís, Y.; Davidenko, N.; García, R. Chitosan/hydroxyapatite-based composites. *Biotecnol. Apl.* **2010**, *27*, 202–210.

32. Pighinelli, L.; Kucharska, M. Properties and structure of microcrystalline chitosan and hydroxyapatite composites. *J. Biomater. Nanobiotechnol.* **2014**, *5*, 128–138. [CrossRef]

33. Wang, Z.; Hu, Q. Preparation and properties of three-dimensional hydroxyapatite/chitosan nanocomposite rods. *Biomed. Mater.* **2010**, *5*, 45007. [CrossRef] [PubMed]

34. Wang, K.; Liao, K.; Goh, K.L. How sensitive is the elasticity of hydroxyapatite-nanoparticle- reinforced chitosan composite to changes in particle concentration and crystallization temperature? *J. Funct. Biomater.* **2015**, *6*, 986–998. [CrossRef] [PubMed]

35. Tishchenko, G.; Bleha, M. Diffusion permeability of hybrid chitosan/polyhedral oligomeric silsesquioxanes (POSS) membranes to amino acids. *J. Membr. Sci.* **2005**, *248*, 45–51. [CrossRef]

36. Xu, D.; Loo, L.S.; Wang, K. Pervaporation performance of novel chitosan-POSS hybrid membranes: Effects of POSS and operating conditions. *J. Polym. Sci. Part B* **2010**, *48*, 2185–2192. [CrossRef]

37. Reno, F.; Carniato, F.; Rizzi, M.; Marchese, L.; Laus, M.; Antonioli, D.; Ren, F. POSS/gelatin-polyglutamic acid hydrogel composites: Preparation, biological and mechanical characterization. *J. Appl. Polym. Sci.* **2013**, 699–706. [CrossRef]

38. Zhang, W.; Camino, G.; Yang, R. Polymer/polyhedral oligomeric silsesquioxane (POSS) nanocomposites: An overview of fire retardance. *Prog. Polym. Sci.* **2017**, *67*, 77–125. [CrossRef]

39. Blanco, I.; Bottino, F.A.; Cicala, G.; Latteri, A.; Recca, A. Synthesis and Characterization of Differently Substituted Phenyl Hepta Isobutyl-Polyhedral Oligomeric Silsesquioxane/Polystyrene Nanocomposites. *Polym. Compos.* **2014**, *35*, 151–157. [CrossRef]

40. Andrade, R.J.; Weinrich, Z.N.; Ferreira, C.I.; Schiraldi, D.A.; Maia, J.M. Optimization of Melt Blending Process of Nylon 6-POSS: Improving mechanical properties of spun fibers. *Polym. Eng. Sci.* **2015**, *55*, 1580–1587. [CrossRef]

41. Raftopoulos, K.N.; Pielichowski, K. Segmental dynamics in hybrid polymer/POSS nanomaterials. *Prog. Polym. Sci.* **2016**, *52*, 136–187. [CrossRef]

42. Lee, S.; Park, S.; Kim, Y. Effect of the concentration of sodium acetate (SA) on crosslinking of chitosan fiber by epichlorohydrin (ECH) in a wet spinning system. *Carbohydr. Polym.* **2007**, *70*, 53–60. [CrossRef]

43. Wang, K.; Loo, L.S.; Goh, K.L. A facile method for processing lignin reinforced chitosan biopolymer microfibres: Optimising the fibre mechanical properties through lignin type and concentration. *Mater. Res. Express* **2016**, *3*. [CrossRef]

44. Weibull, W. A statistical distribution function of wide applicability. *ASME J. Appl. Mech.* **1951**, *18*, 293–297.

45. O'Brien, F.J.; Harley, B.A.; Yannas, I.V.; Gibson, L.J. The effect of pore size on cell adhesion in collagen-GAG scaffolds. *Biomaterials* **2005**, *26*, 433–441. [CrossRef] [PubMed]

46. O'Brien, F.J. Biomaterials & scaffolds for tissue engineering. *Mater. Today* **2011**, *14*, 88–95.

47. Ramshaw, J.A.M.; Werkmeister, J.A.; Dumsday, G.J. Emerging directions for biomedical materials Bioengineered collagens. *Bioengineered* **2017**, *5*, 227–233. [CrossRef] [PubMed]

48. Bhattarai, N.; Edmondson, D.; Veiseh, O.; Matsen, F.A.; Zhang, M. Electrospun chitosan-based nanofibers and their cellular compatibility. *Biomaterials* **2005**, *26*, 6176–6184. [CrossRef] [PubMed]

49. Goh, K.L. *Discontinuous-Fibre Reinforced Composites: Fundamentals of Stress Transfer and Fracture Mechanics*; Springer: London, UK, 2017. [CrossRef]

50. Sadat-Shojai, M.; Khorasani, M.-T.; Jamshidi, A. Hydrothermal processing of hydroxyapatite nanoparticles—A Taguchi experimental design approach. *J. Cryst. Growth* **2012**, *361*, 73–84. [CrossRef]

51. Sadat-Shojai, M.; Khorasani, M.-T.; Dinpanah-Khoshdargi, E.; Jamshidi, A. Synthesis methods for nanosized hydroxyapatite with diverse structures. *Acta Biomater.* **2013**, *9*, 7591–7621. [CrossRef] [PubMed]

52. Notin, L.; Viton, C.; David, L.; Alcouffe, P.; Rochas, C.; Domard, A. Morphology and mechanical properties of chitosan fibers obtained by gel-spinning: Influence of the dry-jet-stretching step and ageing. *Acta Biomater.* **2006**, *2*, 387–402. [CrossRef] [PubMed]

53. Denkba, E.B.; Seyyal, M.; Piskin, E. ImplanTable 5-fluorouracil loaded chitosan scaffolds prepared by wet spinning. *J. Membr. Sci.* **2000**, *172*, 33–38. [CrossRef]

54. Chen, Z.G.; Wang, P.W.; Wei, B.; Mo, X.M.; Cui, F.Z. Electrospun collagen–chitosan nanofiber: A biomimetic extracellular matrix for endothelial cell and smooth muscle cell. *Acta Biomater.* **2010**, *6*, 372–382. [CrossRef] [PubMed]

55. Kumar, R.; Prakash, K.H.; Cheang, P.; Khor, K.A. Temperature driven morphological changes of chemically precipitated hydroxyapatite nanoparticles. *Langmuir* **2004**, *8*, 5196–5200. [CrossRef]

56. Banerjee, A.; Bandyopadhyay, A.; Bose, S. Hydroxyapatite nanopowders: Synthesis, densification and cell–materials interaction. *Mater. Sci. Eng. C* **2007**, *27*, 729–735. [CrossRef]

57. Strachota, A.; Tishchenko, G.; Matejka, L.; Bleha, M. Chitosan–oligo(silsesquioxane) blend membranes: Preparation, morphology, and diffusion permeability. *J. Inorg. Organomet. Polym.* **2002**, *11*, 165–182. [CrossRef]

58. Kopesky, E.T.; McKinley, G.H.; Cohen, R.E. Toughened poly(methyl methacrylate) nanocomposites by incorporating polyhedral oligomeric silsesquioxanes. *Polymer* **2006**, *47*, 299–309. [CrossRef]

59. Ayandele, E.; Sarkar, B.; Alexandridis, P. Polyhedral oligomeric silsesquioxane (POSS)-containing polymer nanocomosites. *Nanomaterials* **2012**, *2*, 445–475. [CrossRef] [PubMed]

60. Motskin, M.; Wright, D.M.; Muller, K.; Kyle, N.; Gard, T.G.; Porter, A.E.; Skepper, J.N. Biomaterials hydroxyapatite nano and microparticles: Correlation of particle properties with cytotoxicity and biostability. *Biomaterials* **2009**, *30*, 3307–3317. [CrossRef] [PubMed]

61. Goh, K.L.; Aspden, R.M.; Mathias, K.J.; Hukins, D.W.L. Finite-element analysis of the effect of material properties and fibre shape on stresses in an elastic fibre embedded in an elastic matrix in a fibre-composite material. *Proc. R. Soc. Lond. A* **2004**, 2339–2352. [CrossRef]

62. Goh, K.L.; Aspden, R.M.; Mathias, K.J.; Hukins, D.W.L. Effect of fibre shape on the stresses within fibres in fibre-reinforced composite materials. *Proc. R. Soc. Lond. A* **1999**, *455*, 3351–3361. [CrossRef]

63. Goh, K.L.; Aspden, R.M.; Hukins, D.W.L. Review: Finite element analysis of stress transfer in short-fibre composite materials. *Compos. Sci. Technol.* **2004**, *64*, 1091–1100. [CrossRef]

64. Fu, S.; Xu, G.; Mai, Y. On the elastic modulus of hybrid particle/short-fiber/polymer composites. *Compos. Part B* **2002**, *33*, 291–299. [CrossRef]

65. Venkateshwaran, N.; Elayaperumal, A.; Sathiya, G.K. Prediction of tensile properties of hybrid-natural fiber composites. *Compos. Part B* **2012**, *43*, 793–796. [CrossRef]

66. Wang, K.; Wu, J.; Ye, L.; Zeng, H. Mechanical properties and toughening mechanisms of polypropylene/barium sulfate composites. *Compos. Part A* **2003**, *34*, 1199–1205. [CrossRef]

67. Estili, M.; Sakka, Y. Recent advances in understanding the reinforcing ability and mechanism of carbon nanotubes in ceramic matrix composites. *Sci. Technol. Adv. Mater.* **2014**, *15*, 64902. [CrossRef] [PubMed]

68. Wang, H.W.; Zhou, H.W.; Peng, R.D.; Mishnaevsky, L. Nanoreinforced polymer composites: 3D FEM modeling with effective interface concept. *Compos. Sci. Technol.* **2011**, *71*, 980–988. [CrossRef]

69. Aranaz, I.; Martínez-Campos, E.; Moreno-Vicente, C.; Civantos, A.; García-Arguelles, S.; del Monte, F. Macroporous calcium phosphate/chitosan composites prepared via unidirectional ice segregation and subsequent freeze-drying. *Materials* **2017**, *10*, 516. [CrossRef] [PubMed]

70. Irvine, S.A.; Venkatraman, S.S. Bioprinting and differentiation of stem cells. *Molecules* **2016**, *21*, 1188. [CrossRef] [PubMed]

71. Lange, F.F.; Radford, K.C. Fracture energy of an epoxy composite system. *J. Mater. Sci.* **1971**, *6*, 1197–1203. [CrossRef]

72. Sadat-Shojai, M.; Atai, M.; Nodehi, A.; Khanlar, L.N. Hydroxyapatite nanorods as novel fillers for improving the properties of dental adhesives: Synthesis and application. *Dent. Mater.* **2010**, *26*, 471–482. [CrossRef] [PubMed]

© 2017 by the authors. Licensee MDPI, Basel, Switzerland. This article is an open access article distributed under the terms and conditions of the Creative Commons Attribution (CC BY) license (http://creativecommons.org/licenses/by/4.0/).

Journal of
composites science

MDPI

Article

Computational Study of the Effects of Processing Parameters on the Nonlinear Elastoplastic Behavior of Polymer Nanoclay Composites

Arifur Rahman and Xiang-Fa Wu *

Department of Mechanical Engineering, North Dakota State University, Fargo, ND 58108, USA;
Arifur.Rahman@ndsu.edu
* Correspondence: xiangfa.wu@ndsu.edu; Fax: +1-701-231-8913

Received: 29 September 2017; Accepted: 7 December 2017; Published: 9 December 2017

Abstract: Processing parameters (e.g., exfoliation extent and volume fraction) of clay particles in polymeric resins play a crucial role in the mechanical properties of polymer nanoclay composites (PNCs). This paper is aimed to investigate the effects of clay aspect ratio and volume fraction on the global mechanical properties (e.g., effective stiffness, yield strength, and ultimate tensile strength) of PNCs. During the process, computational micromechanics models are adopted to simulate the nonlinear elastoplastic behavior of the PNCs of varying clay particle volume fractions and aspect ratios subjected to uniaxial tension. A representative volume element (RVE) of the PNCs is employed for the finite-element-method (FEM) based computational simulations. The polymeric matrix is treated as an idealized elastoplastic solid with isotropic hardening behavior, and the clay particles are treated as stiff elastic platelets distributed evenly in the stack and stagger configurations in the matrix. Seven volume fractions (V_f = 0.5%, 1%, 2%, 5%, 7.5%, 10%, and 15%) and seven aspect ratios (the ratio of platelet length over thickness ρ = 1, 2, 5, 10, 20, 50 and 100) of the reinforcing clay particles are utilized. Numerical experiments show that the effective modulus of the PNCs at small strains increases with the increase of either the clay volume fraction or the platelet aspect ratio largely following those predicted by classic micromechanics models. However, at the low particle aspect ratios (e.g., ρ = 1, 2, 5 and 10), the ultimate tensile strength of the clay composite is nearly independent of the clay volume fraction up to 5% in the present study, i.e., the polymeric matrix governs the PNC strength; at the large particle aspect ratios (e.g., ρ = 20 and 50), the ultimate tensile strength is significantly enhanced with growing clay volume fraction higher than 5% and reaches ~150% of that of the polymeric matrix at ρ = 50 and V_f = 10%. A comparative study is conducted for stack and stagger models for the prediction of the mechanical properties of PNCs. It shows that the stack model predicts slightly larger values of the effective stiffness and tensile strength than the stagger model. The numerical study shows that a large platelet aspect ratio through full exfoliation of the clay particles in matrix is crucial to achieving the preferable mechanical properties of PNCs as evidenced in experiments. The present results can be utilized to quantitatively explain the mechanical properties of clay particle-reinforced composites and PNCs within the framework of classic micromechanics, and provide guidelines for computer-aided nanocomposites design for processing property-tailorable PNCs.

Keywords: polymer nanoclay composites (PNCs); mechanical behavior; strength and stiffness; scaling analysis; elastoplastic; finite element analysis (FEA)

---

1. Introduction

Nanocomposites made of polymeric matrices reinforced with intercalated or exfoliated clay nanoparticles have become a focus of research in polymer composites in the last two decades after the seminal research initiated successfully by the Toyota research group in 1980s [1,2]. To date, substantial experimental investigations have shown that layered-silicate clay particles can be exfoliated into single nanosized platelets through ion exchange and can be well distributed in polymer melts or solutions as a nanoreinforcing phase in enhancing the mechanical and other physical properties of the resulting polymer nanoclay composites (PNCs). The most attractive evidence as demonstrated by numerous researchers worldwide is that layered silicate nanofillers can tremendously increase the tensile modulus, flexural stiffness and the tensile strength of PNCs at a filler weight content of only a few percent. Such preferable experimental observations are due mainly to the fact that the tensile modulus and strength of well aligned, exfoliated nanoclay platelets are close to the theoretical values of their perfect crystalline counterparts without defects (dislocations) and orders higher than those of the polymeric resins, as well as the high interfacial bonding strength (shear strength) between the clay nano platelets and the polymeric resins that exhaustively exploits the toughening effect of the clay nano platelets. For instance, nylon-6 nanocomposites prepared through intercalative ring-opening polymerization of ε-caprolacetam modified montmorillonite to form fully exfoliated nano platelets can double the tensile modulus of the virgin nylon matrix at the filler weight content at 4.1% [3]; however, such giant modulus improvement cannot be achieved when the clay particles are dispersed in polymeric matrix to form intercalated polymer microcomposites [4]. Meanwhile, fully exfoliated clay nanofillers at a weight content of ~5% can also enhance the tensile strength of the nanocomposites up to 50% as demonstrated in nylon nanocomposites that were prepared through in situ intercalative polymerization of ε-ecaprolactam in protonated aminododecanoic acid modified montmorillonite [4]. Besides, increasing experimental data have indicated that PNCs are capable of having desirable thermal stability, flame retardancy, and gas barrier properties that substantially broaden the applications of conventional polymer composites [4–10] in various industrial sectors. Among those, the most recent research investigated the effects of nanoclay morphologies (e.g., particle length, aspect ratio, bird nest structure, aggregation, etc. of halloysite and kaolinite nanoclay particles) on the thermal stability, surface wettability, and mechanical properties of biopolymer nanocomposite films for food packaging and other applications [11,12]. The studies showed that the thermal, tensile and surface wetting properties of the pectin resin can be noticeably improved by exfoliated halloysite nanotubes at relatively high weight concentration, which form unique bird nest structures in the pectin matrix of the resulting bionanocomposites. In addition, most recent research on the structural, physical and mechanical properties of nanoclay-reinforced fiber composites can be found in Reference [13].

Along with the enrichment of the database of experimental data of PNCs, remarkable efforts has been devoted to model-based understanding and numerical simulation of the mechanical behavior of PNCs for controllable processing and nanocomposites design. One typical treatment broadly considered by researchers is to consider PNCs as conventional particle-reinforced composites while an effective particle volume (size) is assumed [14]. Such treatment facilitates the utilization of classic micromechanics and numerical methods (e.g., finite element method-FEM) to analyze the mechanical behavior of PNCs [15]. With the assumption that the nanofillers are evenly distributed in the matrix, the effective mechanical properties, mainly the effective moduli, of the resulting nanocomposites can be approximately modeled by adopting Eshelby's equivalent model, self-consistent models of finite-length fillers, Mori-Tanaka type models, bound models, Halpin-Tsai model and its extensions, and modified shear-lag models, among others [15]. By comparison with these classic micromechanics models plus detailed finite element analysis (FEA), Tuchker and Liang concluded that the simple shear-lag model is capable of giving a good estimate of the longitudinal modulus (E_{11}) of short-fiber reinforced composites when the fiber aspect ratio is greater than 10 [15]. Meanwhile, Tsai and Sun [16] developed a modified shear-lag model to investigate the load transfer efficiency in PNCs reinforced with intercalated and fully exfoliated clay particles, and their model was validated by using FEM. In addition, Weon and Sue [17] developed an experimental scheme to tailor the clay orientation and aspect ratio in nylon-6 PNCs through controlled shearing in order to study their effects on the mechanical properties. It was found that the effective modulus and tensile strength of the PNCs decreased with decreasing aspect ratio and alignment extent of the reinforcing clay nano platelets; in contrast, the fracture toughness and ductility of the PNCs increase simultaneously. In the study, the Halpin-Tsai and Mori-Tanaka micromechanics models were employed successfully to examine the dependency of composite moduli upon the orientation and aspect ratio of the clay platelets; the effective filler structure was established through mapping the effective structural parameters of fillers to the structural parameters of conventional fillers within the framework of micromechanics [18]. Similar studies were also conducted by many other researchers. For example, Zhu and Narh [19] performed detailed FEA of the mechanical behavior of aligned clay platelet-reinforced nanocomposites. In their model, the clay particles and the polymeric matrix were all treated as linearly elastic solids and an interlayer was proposed to model the transition region between the polymeric matrix and the fillers. Numerical simulations based on such a three-phase model showed that either decreasing interlayer modulus or increasing interlayer thickness resulted in decreasing effective moduli of the PNCs. In addition, Dai et al. [20] recently introduced an improved stagger model to analyze the stiffness and strength of PNCs reinforced with aligned, exfoliated clay nanoparticles; numerical results are close to those obtained in experimental studies. More recently, Dong and Bhattacharyya [21] performed finite element modeling to determine the effective moduli of polypropylene/organoclay nanocomposites. In the model, representative volume elements (RVEs) of stacking, staggering, and randomly distributed platelets in matrices were considered; their numerical results showed that the interlayer properties had less impact on the effective moduli of the PNCs reinforced with fully exfoliated clay nanoparticles. More detailed reviews on research progress in processing, characterization and modeling of the mechanical properties of PNCs can be found in the recent review papers in this topic, e.g., [4–6,13,21,22] and references therein.

In the above, it can be observed that the majority of the modeling studies reported in the literature was focused on the effective moduli of PNCs by adopting the classic micromechanics models and related computational methods within the range of linearly elastic deformation of the composites. Yet, no systematic study has been conducted on the ultimate tensile strength of PNCs though it is an important material parameter in practical applications of PNCs. In view of processing PNCs, the key processing parameters include the clay particle volume (or weight) fraction, exfoliation extent and platelet orientation in polymeric matrices, which dominate the mechanical properties of the resulting PNCs. Thus, in this study, we initiated a computational scaling study on the effects of processing parameters (i.e., the volume fraction and aspect ratio of the clay particles) on the effective modulus, yield strength, and ultimate tensile strength of PNCs by using a nonlinear FEM model. In the model, the typical stack and stagger RVEs of PNCs were adopted, and the polymeric matrix was treated as idealized elastoplastic solid with isotropic hardening behavior. A family of effective stress-strain diagrams was gained at varying volume fraction and aspect ratio of the clay particles through detailed nonlinear FEA. Dependencies of the effective stiffness, yield strength and ultimate tensile strength upon the processing parameters of the PNCs and relevant mechanisms were explored and discussed. Conclusions of the present study and relevant applications were addressed in the last section of the paper.

2. Problem Statement and Simulation

2.1. Scaling Analysis of the Mechanical Properties of PNCs

Given a PNC system, the processing parameters such as the particle volume (weight) fraction V_f, aspect ratio ρ, defined as the ratio of the platelet length to the thickness in the simplified two dimensional (2D) case, and the platelet orientation angle θ, among others, dominate the mechanical properties (e.g., effective modulus E_e, yield strength σ_{ye}, and ultimate tensile strength σ_{ue}) of the resulting nanocomposite material. In this study, without loss of the generality, we adopted the 2D stack and stagger RVEs, in which all the clay platelets are assumed to be distributed either in stacked or staggered patterns in the polymeric matrix. According to the Buckingham π-theorem [23], the effective modulus E_e, yield strength σ_{ye}, and ultimate tensile strength σ_{ue} of the PNC have the following scaling relationships with respect to the constituent moduli (E_f and E_m), yield strength (σ_{ym}), and Poisson's ratios (υ_f and υ_m) and the volume fraction (V_f) and aspect ratio (ρ) of the clay particles:

$$\frac{E_e}{E_m} = f\left(\frac{E_f}{E_m}, \frac{\sigma_{ym}}{E_m}, \upsilon_f, V_m, V_f, \rho\right) \tag{1}$$

$$\frac{\sigma_{ye}}{E_m} = g\left(\frac{E_f}{E_m}, \frac{\sigma_{ym}}{E_m}, \upsilon_f, V_m, V_f, \rho\right) \tag{2}$$

$$\frac{\sigma_{ue}}{E_m} = h\left(\frac{E_f}{E_m}, \frac{\sigma_{ym}}{E_m}, \upsilon_f, V_m, V_f, \rho\right) \tag{3}$$

where f, g, and h are, respectively, three unknown dimensionless functions with respect to the dimensionless ratios E_f/E_m, σ_{ym}/E_m, and dimensionless numbers υ_f, υ_m, V_f and ρ. In the above, subscript 'e' denotes the effective properties of the PNC; subscript 'f' denotes the properties of clay particles, and subscript 'm' stands for the properties of the polymeric matrix. It needs to be mentioned that in reality, more parameters of a PNC system potentially govern the global mechanical properties of the PNC, such as the interfacial shear strength between the clay nano platelets and polymer resin, the morphology and orientation of the clay particles, among other. Yet, as a preliminary computational approach, the present study focuses only on the dominant parameters within the framework of classic mechanics of composite materials and computational micromechanics.

In typical PNCs, the clay particles are much stiffer and have much higher yield strength and ultimate tensile strength than those of the polymeric resins. Therefore, subjected to external loading, the clay particles are only in the range of elastic deformation till the catastrophic failure of the PNCs. As a reasonable deduction, the yield and tensile strengths of the clay particles can be ignored in the scaling relations (1–3). For the purpose of simplifying the upcoming discussions, the polymers in this study are assumed to be idealized elastoplastic solid, i.e., the yield strength σ_{ym} equals to its ultimate tensile strength σ_{um}. In addition, the yield strength of a PNC (2) σ_{ye} is physically less meaningful since in computational simulation such as FEA, localized tiny plastic deformation always exists in the polymeric matrix near the ends of the clay particles due to stress concentration at the clay particle edge/corner regions with abrupt geometrical change and depends also upon the extent of mesh refinement employed in FEA. However, the global mechanical properties of PNCs such as the tensile strength and fracture toughness are not so sensitive to such localized plastic deformation since the polymers are usually thermoplastic with excellent ductility and plastic deformation.

Besides, the transition region between the clay platelets and the polymeric matrix is not taken into account in the current study due to the following two considerations. First, physically, it is still rather difficult to identify the exact thickness of such interlayers based on available experimental data in the literature though many researchers have artificially introduced this ad hoc region to show the size effect in the mechanical properties of nanocomposites. Second, some recent simulations such as those provided by Dong and Bhattacharyya [21] showed less impact of the interlayers on the mechanical properties of PNCs, which might be related to the small volume fraction of the interlayers and the clay particles in the model. Thus, introduction of the interlayers to modeling the mechanical properties of PNCs is still a controversial topic and needs more in-depth experimental evidence and physical justification. Thus, in the present study, we mainly focused on the scaling properties of the effective modulus (1) and the ultimate tensile strength (2) of the PNCs with respect to the two major processing parameters: the clay-particle volume fraction V_f and geometrical aspect ratio ρ. A detailed 2D nonlinear FEA was conducted for this purpose.

2.2. FEM-Based Computational Micromechanics Simulation

In processing a PNC, clay particles can be potentially distributed in the polymeric matrix in three different extents, i.e., the conventional micron level particles, intercalated particle and exfoliated nano platelets, as illustrated in Figure 1. The latter two are the general states to achieve PNCs with preferable mechanical properties.

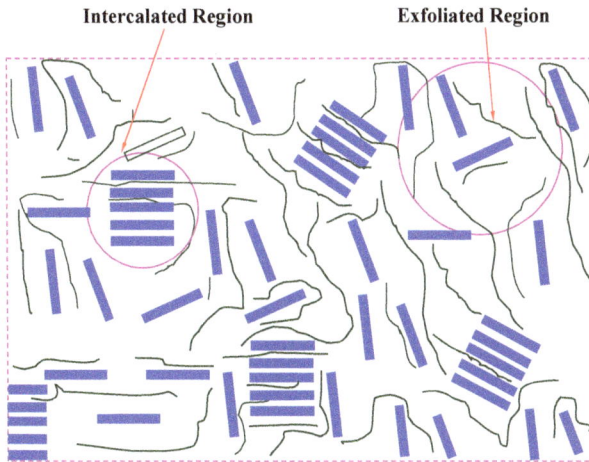

Figure 1. Schematic of intercalated and exfoliated clay particles in a polymeric matrix.

To investigate the scaling properties of PNCs with varying clay-particle volume fraction V_f and aspect ratio ρ, we considered the typical stacking and staggering distributions of identical clay particles in the polymeric matrix. The RVEs to be extracted for FEM simulations are based on the periodical and symmetrical conditions of the PNCs subjected to axial tension along the particle axis, as shown in Figures 2 and 3, respectively.

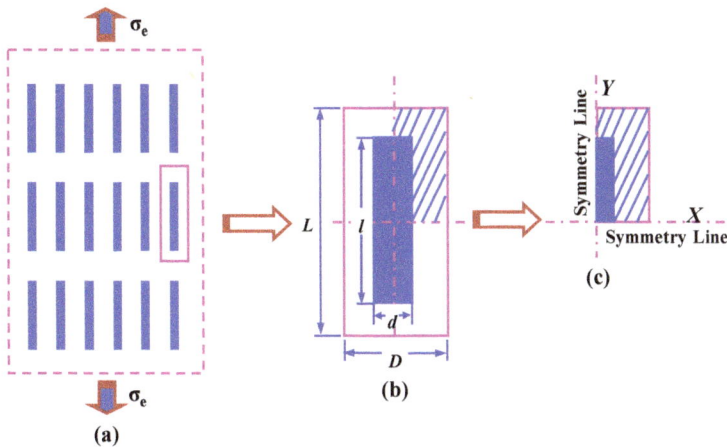

Figure 2. Representative volume element (RVE) of the stack model used for the present computational scaling analysis. (**a**) Idealized identical stacking clay platelets; (**b**) a typical RVE; and (**c**) a quarter RVE for efficient simulation.

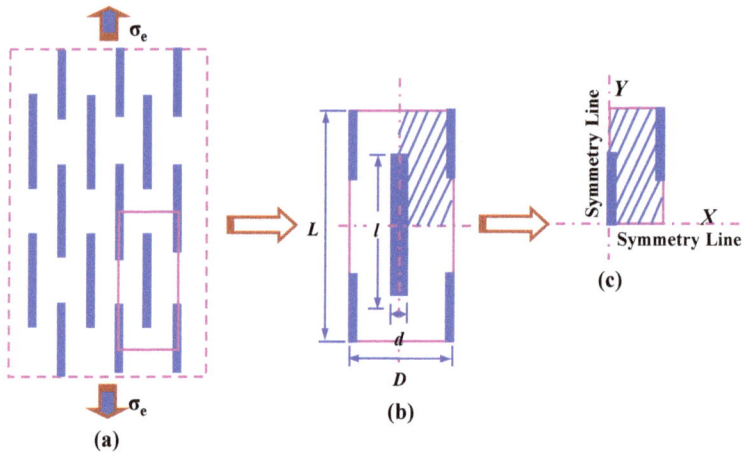

Figure 3. Representative volume element (RVE) of the stagger model used for the present computational scaling analysis. (**a**) Idealized identical staggering clay platelets; (**b**) a typical RVE; and (**c**) a quarter RVE for efficient simulation.

During the computational simulation, a wide range of clay-particle aspect ratios (ρ = 1, 2, 5, 10, 20, 50 and 100) and volume fractions (V_f = 0.5%, 1%, 2%, 5%, 7.5%, and 10%) were utilized. Additionally, the stack and stagger models with the filler volume fraction 15% and aspect ratio 100 have been also considered for the purpose of comparative study, to examine their effects on the ultimate tensile strength and effective modulus of the resulting PNCs. The aspect ratio ρ and volume fraction V_f can be obtained by adjusting the RVE dimensions: L, D, l, and d as

$$\rho = \frac{l}{d} \text{ and } V_f = \frac{ld}{LD}. \tag{4}$$

The mechanical properties of the polymeric matrix and clay particles were selected close to those reported in the literature [4,6]. The polymeric matrix was modeled as an idealized elastoplastic solid with the yield strength σ_{ym} = 79.0 MPa, Young's modulus E_m = 2.75 GPa, and Poisson's ratio v_m = 0.41; the clay particles were modeled as linearly elastic solid with Young's modulus E_f = 178 GPa and Poisson's ratio v_f = 0.28.

2D nonlinear FEA was conducted to determine the entire stress field of the quarter RVEs (see Figures 2c and 3c) of the PNCs subjected to axial stretching. During the process, a plane-strain 4-node 182 element offered by ANSYS™ was adopted; symmetrical boundary conditions were enforced at the horizontal and vertical symmetric axes; forced displacement constraints were triggered to maintain the constant horizontal displacements at the right vertical edges; no debonding was assumed between the clay particle and the matrix. The nonlinear FEA was executed under the condition of small displacement; the Newton-Raphson algorithm was selected for the nonlinear numerical iterations. The convergence criterion, equation solver, and nonlinear options were set as program defaults. After each simulation, the nodal forces at the top edge were extracted and recorded at several sampling displacements (i.e., effective tensile strains) until the effective strain equaled three times the yield strain of the polymeric matrix; the corresponding effective tensile stresses were calculated manually by dividing the sum of the edge nodal forces by the width of the RVE. The numerical experiments showed that the above nonlinear FEA can give reliable effective stress-stain diagrams of the PNCs at varying clay-particle aspect ratio ρ and volume fraction V_f within this consideration. Detailed numerical results will be discussed in Section 3.

3. Results and Discussion

In the present computational micromechanics scaling study, FEM-based nonlinear analysis was conducted to cover a wide range of clay-particle aspect ratios (ρ = 1, 2, 5, 10, 20, 50 and 100) and volume fractions (V_f = 0.5%, 1%, 2%, 5%, 7.5%, and 10%) for both the stack and stagger models. The effective moduli predicted by the two idealized clay-particle arrangements (i.e., stacking and staggering) are nearly the same, especially at the low volume fraction of dilute clay particles in the polymeric matrix. This is the trivial conclusion that can be easily drawn in linearly elastic micromechanics as the effective moduli of PNCs of aligned clay particles are independent of either stacking or staggering arrangement in the dilute clay particle state, and the small variation between the two models are possibly induced by the difference of localized plastic deformations and a small deviation of the clay particle interaction in the two micromechanics models. Compared to the effective moduli, the effective ultimate tensile strengths of the PNCs have a little noticeable variation between the stack and stagger models. However, the general varying trends of the effective moduli and ultimate tensile strengths based on the two computational micromechanics models are consistent and will be discussed further. Hereafter, only the results based on the stack model are shown though a comparative study will also be presented later on. In the case of the stack model, variation of the effective elastic modulus E_e and effective ultimate tensile strength σ_{ue} with the varying clay-particle volume fraction V_f at several aspect ratios are shown in Figures 4 and 5, respectively. During the data reduction process, the effective moduli in Figure 4 were extracted from the slope of the initial linear elastic region of the effective tensile stress-strain diagrams (σ_e-ε_e) of all the computational cases as shown in Figures 6–11. These results indicate a very good linear relationship between the effective tensile stress σ_e and the effective tensile strain ε_e as expected. From Figures 6–11, it can be observed that when the value of effective tensile strain ε_e is low, the PNCs behave as linearly elastic solids until the effective tensile strain ε_e is close to the yield strain of the polymeric matrix. The effective elastic moduli E_e extracted from the present numerical experiments (see Figure 4) are close to those available in the literature [11,12]. In addition, it can also be found from Figures 6–11 that the PNCs exhibit slightly softening behavior prior to the arrival of the yield strain of the polymeric matrix. This phenomenon can be attributed to the localized plastic deformation near the ends of clay particles where stress concentration exists due to the sharp edges/corners of the clay particles. In reality, at the nanoscale, the physical/chemical properties of the polymer chains close to the clay platelets are much more complicated than the present material model and are also noticeably different from the polymers in bulk state. Such unusual properties of the interfacial polymers at the nanoscale are beyond the scope of this computational study and are still under intensive investigation.

In addition, at the low clay-particle volume fraction (V_f < 2%) or low aspect ratio (ρ = 1, 2, 5, and 10) based on the present computational study, no obvious improvement of the effective ultimate tensile strength σ_{ult} is detected as shown in Figure 5. This characteristic can be also examined from the σ_e-ε_e diagrams given in Figures 6–11. Such an observation is also qualitatively in agreement with the recent experimental results of the mechanical properties of PNCs as reported in the literature [4,6]. In these cases, the values of effective ultimate tensile strength σ_{ult} are still dominated by the yield strength of the polymeric matrix; the contribution of the load-carrying capacity of clay particles to the nanocomposites is negligible. Such an observation can be understood as follows. Physically, different from the effective stiffness which is governed by the averaging effect of the mechanical properties of the constituents forming the PNCs, the dilute clay nanoparticles in the polymer matrix are unable to form load-transferring bridges and therefore unable to noticeably strengthen the resulting PNCs. Therefore, dilute clay nanoparticles at low volume concentration have less impact on the hardening of the polymeric matrix. However, the effective ultimate tensile strength σ_{ult} of the PNCs is significantly enhanced in the cases of V_f = 5% and 10% and ρ = 50 as shown in Figures 10 and 11. In the case of V_f = 5% and ρ = 20 (Figure 10), the improvement of the effective ultimate tensile strength σ_{ult} of the PNC reaches ~60% on the basis of the yield strength of the virgin polymeric matrix; this increment is close to those achieved experimentally in Nylon-6 and silicone rubber nanocomposites reinforced with montmorillonite clay particles [4]. Moreover, in the case of $V_f = 10\%$ and $\rho = 50$, the present simulation predicts the increment of the effective ultimate tensile strength σ_{ult} of the PNCs up to ~150%. This value is much higher than the ones reported in the literature. In realistic PNCs, the polymer/clay interphase, interfacial shear failure (sliding/pull-out failure), filler waviness, and orientation will noticeably decrease this value. In the above two cases, the clay particles with an aspect ratio ρ higher than 10 apparently increase the load-carrying capacity of the resulting PNCs.

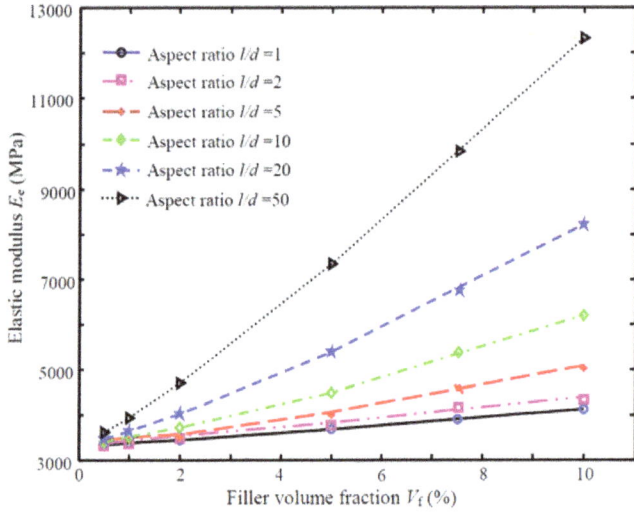

Figure 4. Variation of the effective elastic modulus E_e with the filler volume fraction V_f.

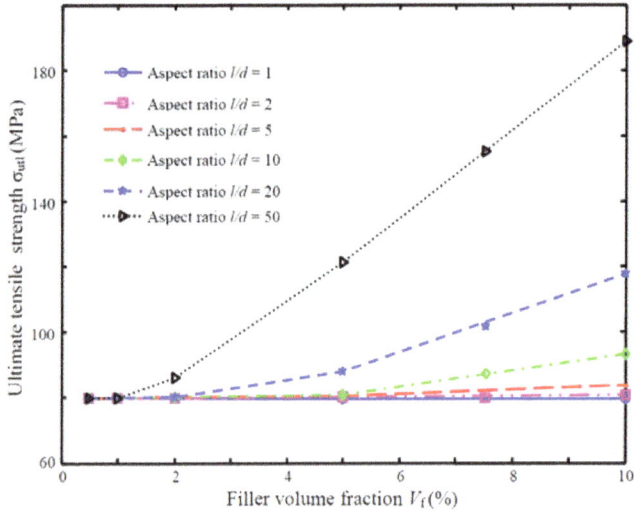

Figure 5. Variation of the effective ultimate tensile strength σ_{utl} with the filler volume fraction V_f.

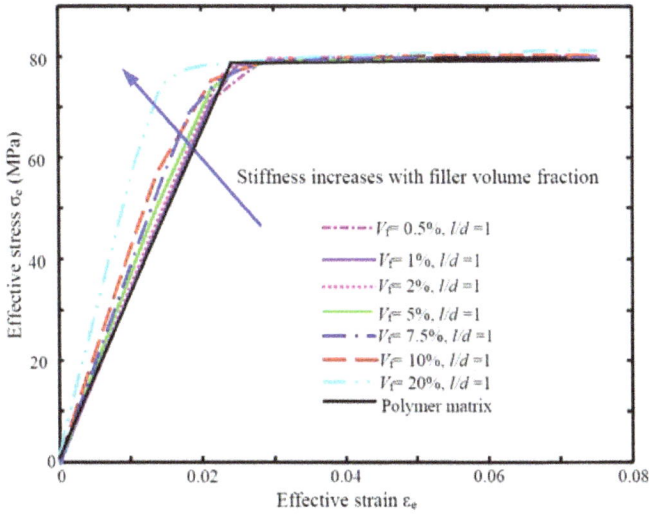

Figure 6. Effective stress-strain σ_e-ε_e diagram of the PNCs with the clay-particle aspect ratio $\rho = 1$ (plane-strain, case I).

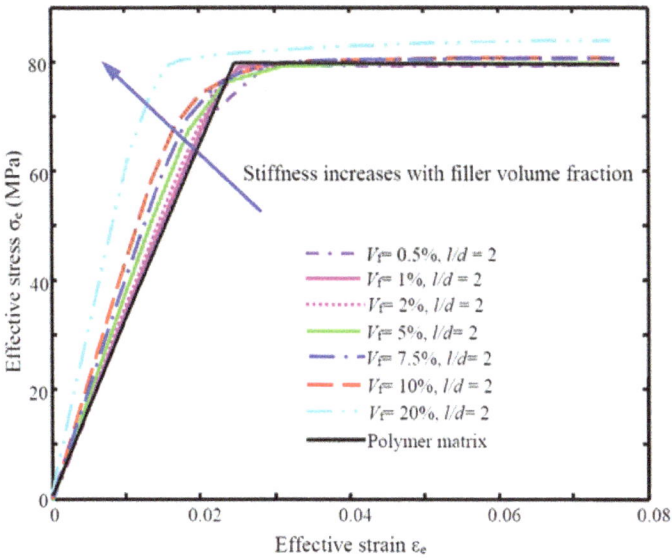

Figure 7. Effective stress-strain diagram σ_e-ε_e of the PNCs with the clay-particle aspect ratio $\rho = 2$ (plane-strain, case I).

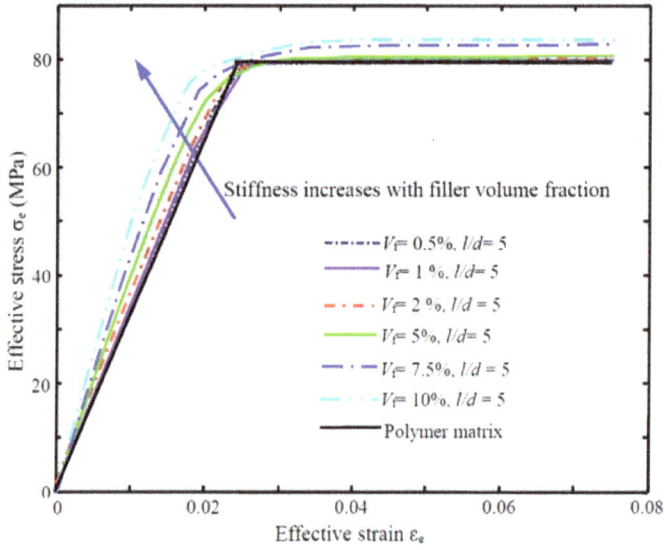

Figure 8. Effective stress-strain diagram σ_e-ε_e of the PNCs with the clay-particle aspect ratio $\rho = 5$ (plane-strain, case I).

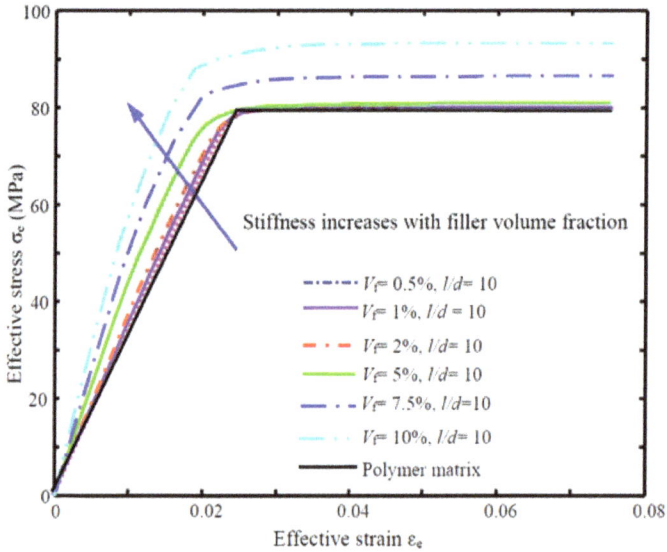

Figure 9. Effective stress-strain diagram σ_e-ε_e of the PNCs with the clay-particle aspect ratio $\rho = 10$ (plane-stain, case I).

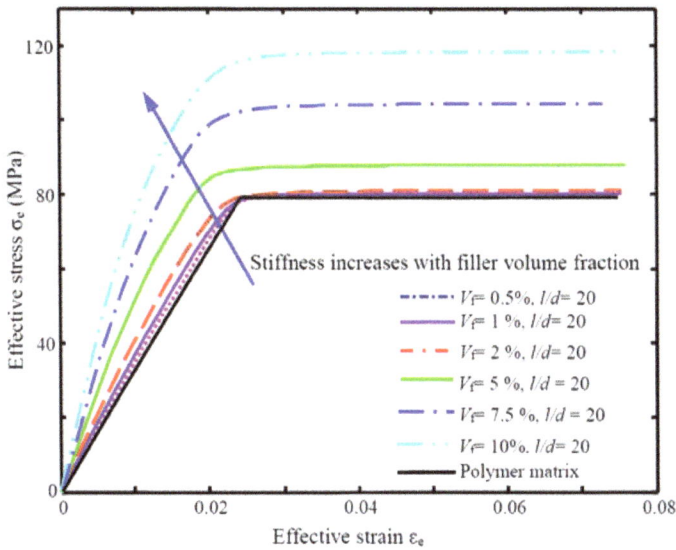

Figure 10. Effective stress-strain diagram σ_e-ε_e of the PNCs with the clay-particle aspect ratio $\rho = 20$ (plane-strain, case I).

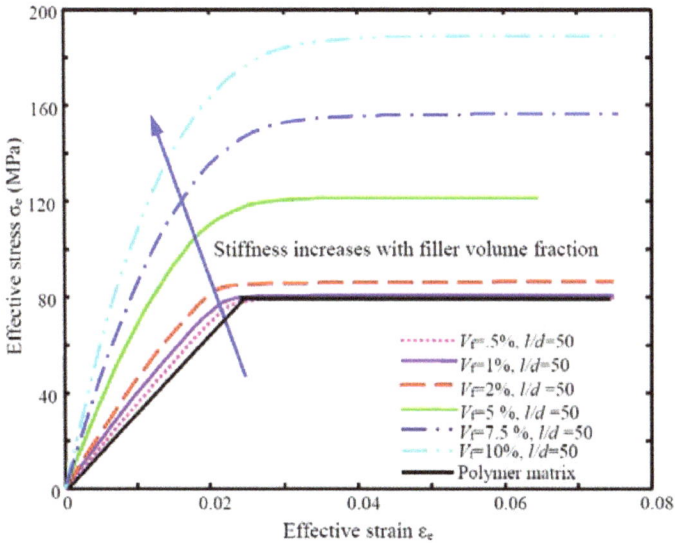

Figure 11. Effective stress-strain diagram σ_e-ε_e of the PNCs with the clay-particle aspect ratio $\rho = 50$ (plane-strain, case I).

Furthermore, the FEA results indicate that at the low particle aspect ratios (e.g., $\rho = 1, 2, 5$ and 10), the effective ultimate tensile strengths σ_{ult} of the PNCs based on the two models are nearly independent of the clay volume fraction V_f up to 5%, i.e., the polymeric matrix governs the tensile strength of the PNCs. This observed phenomenon is due to the fact that the relatively dilute clay nanoparticles in polymeric matrix are unable to form load-transferring bridges at the low aspect ratios of the clay particles to strengthen the resulting PNCs. Besides, by examining the results obtained in the cases of the filler aspect ratio ρ at 10, 20, 50 and 100 and the filler volume fraction V_f at 0.5%, 1%, 2%, 5%, 10% and 15%, respectively, it can be found from Figures 12–15 that when the effective tensile strain ε_e is low, the PNCs behave as linearly elastic solids until the effective strain ε_e close to the yield strain of the polymeric matrix and the stack and stagger models both predict very similar effective tensile moduli. In addition, the PNCs exert slightly softening behavior prior to the arrival of the yield strain of the polymeric matrix due to stress-concentration induced matrix (resin) yielding near the filler ends. In addition, the percent increase of the effective modulus E_e for both the stack and stagger models as shown in Figure 16 indicates that for all the aspect ratios, the computational predictions based on the stack and stagger models are close, especially at the dilute clay particle state, as confirmed by linearly elastic micromechanics. Yet, a small variation of the effective tensile modulus between the stack and stagger models can be detected, which is due to the variation of clay particle interference between the two models. Moreover, Figure 17 shows the comparative effective ultimate tensile strength σ_{ult} of PNCs predicted by the stack and stagger models. It can be observed that the trend of ultimate tensile strength σ_{ult} increase is the same for the two models; however, the deviation of the predicted effective tensile strength increases with increasing aspect ratio ρ for the two models, and the stagger model predicts relatively lower values of the effective tensile strength than the stack model does for all the computational cases under the present consideration. By comparing the clay-particle arrangements in the two models, it can be concluded that the clay nanoparticles in stagger model are arranged more uniform than those in the stack model at a given filler volume fraction. Thus, subjected to the same effective tensile strain, the stack model generated more uneven stress field, or a higher stress gradient, than the stagger model, i.e., a higher ultimate tensile strength by the stack model. Such an observation also indicates that the ultimate tensile strength is more sensitive to the clay nanoparticle arrangement in the model, which can be useful to process strength-controllable PNCs.

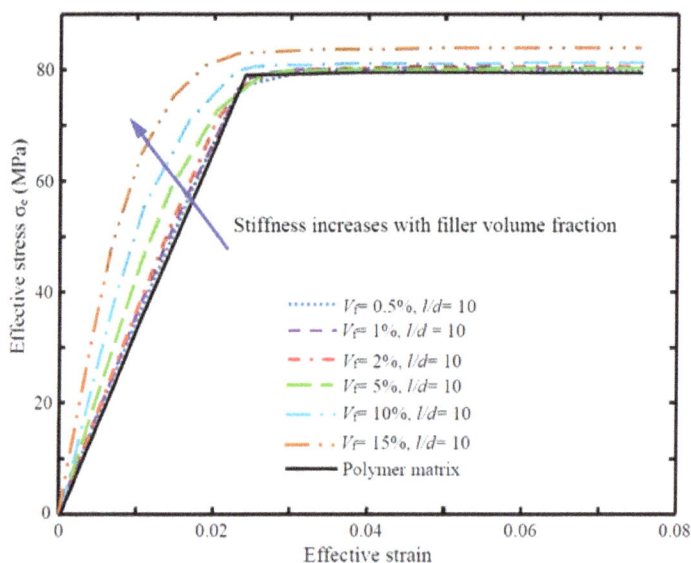

Figure 12. Effective stress-strain diagram σ_e-ε_e of the PNCs with the clay-particle aspect ratio $\rho = 10$ (plane-stain, case II).

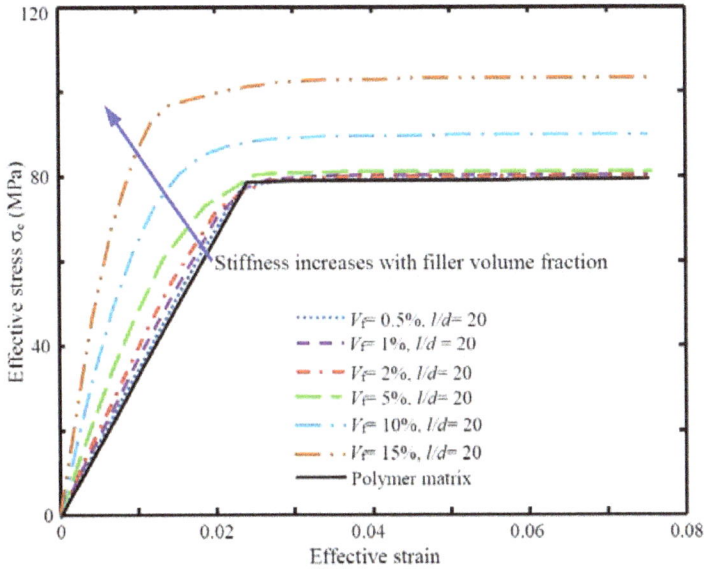

Figure 13. Effective stress-strain diagram σ_e-ε_e of the PNCs with the clay-particle aspect ratio $\rho = 20$ (plane-strain; case II).

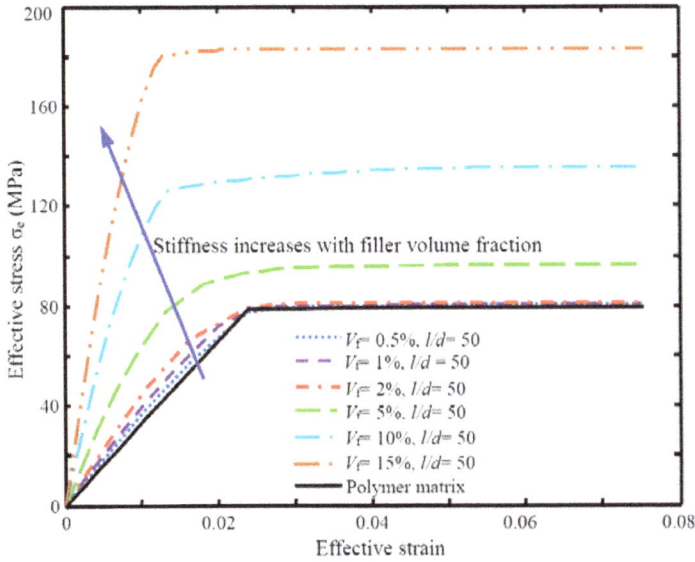

Figure 14. Effective stress-strain diagram σ_e-ε_e of the PNCs with the clay-particle aspect ratio $\rho = 50$ (plane-strain, case II).

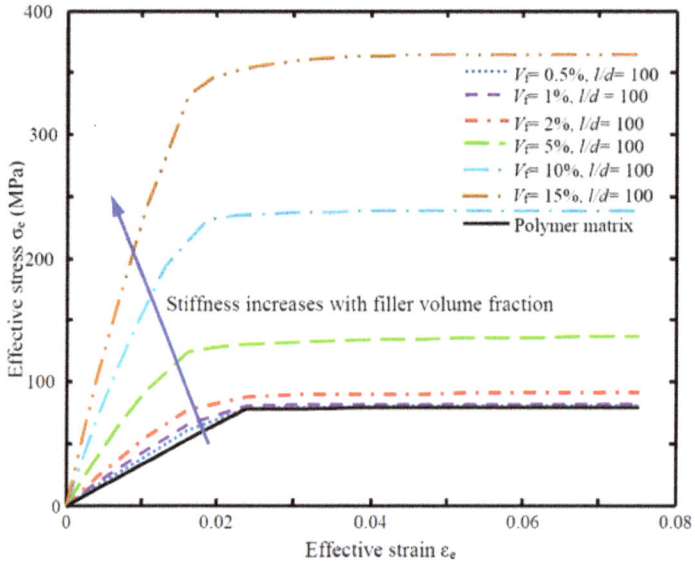

Figure 15. Effective stress-strain diagram σ_e-ε_e of the PNCs with the clay-particle aspect ratio $\rho = 100$ (plane-strain, case II).

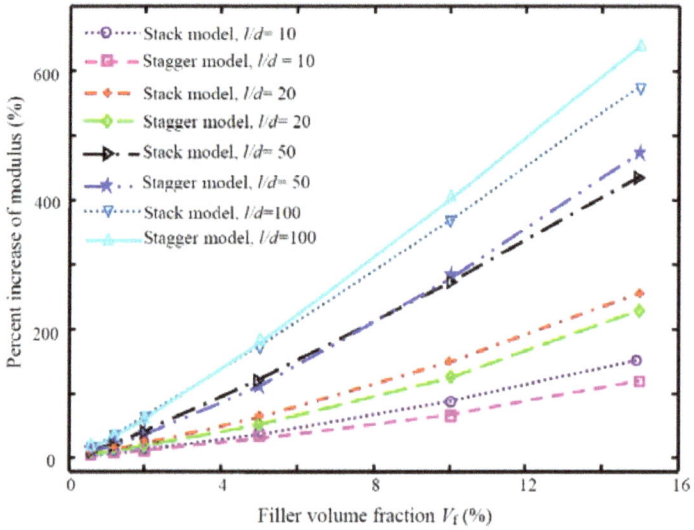

Figure 16. Variation of the percent increase of effective modulus (compared to matrix) with the filler volume fraction V_f for different clay-particle aspect ratios $\rho = 10, 20, 50$ and 100.

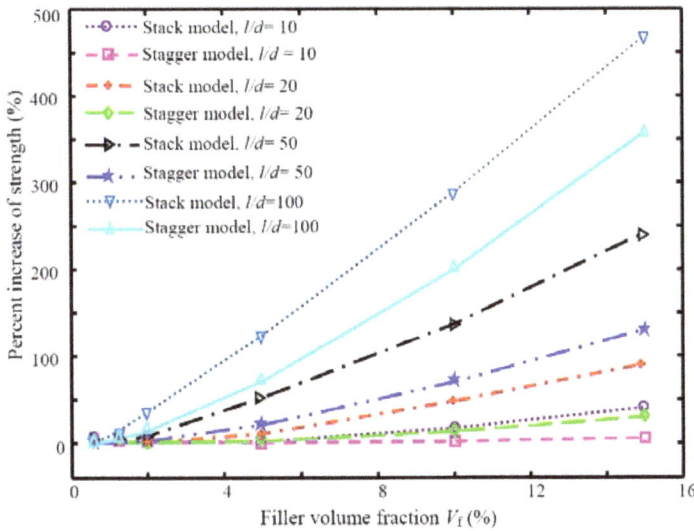

Figure 17. Variation of the percent increase of tensile strength (compared to matrix) with the filler volume fraction V_f for different clay-particle aspect ratios $\rho = 10, 20, 50$ and 100.

It needs to be mentioned that in reality, the aspect ratio of fully exfoliated clay platelets is much larger than the values studied in this work. Yet, not well aligned clay platelets in polymeric matrices will substantially weaken the potential improvement of the effective tensile stiffness E_e and ultimate tensile strength σ_{eu} of the realistic PNCs in a specific orientation. This implies the tailorability of the effective tensile moduli and ultimate tensile strength of PNCs via optimal and controlled clay-particle exfoliation and orientation. To date, efficient processing techniques are still desired to effectively achieve controlled platelet orientation in polymeric matrix. Though the present study was based on the simple stack and stagger models, the computational results provide the insight to understand the scaling mechanical properties of PNCs. Such computational models can be further enhanced to involve other processing parameters such as platelet orientation, waviness and extent of intercalated clay particles for predicting the mechanical behavior of PNCs in controlled fabrication.

4. Concluding Remarks

Computational stack and stagger micromechanics models were formulated and integrated into 2D FEA of PNCs for examining the effects of two key processing parameters (i.e., the clay-particle volume fraction V_f and aspect ratio ρ) on the mechanical behavior of PNCs. The present computational scaling studies have shown that the clay-particle aspect ratio ρ plays a crucial role in enhancing the ultimate tensile strength of PNCs. Specifically, the aspect ratio $\rho > 50$ is preferable to achieve significantly improved mechanical properties of PNCs. The stack and stagger models predicted the close effective moduli and slightly different ultimate tensile strengths due to the model sensitivity in the case of plastic failure of the PNCs. As a matter of fact, completely stacked and staggered clay platelets are two limiting aligned cases (best alignment) of PNCs reinforced with fully exfoliated clay nanoparticles, which predict the effective mechanical properties that can be considered as the theoretical bounds useful to guide experimental studies. The present study was based on classic computational micromechanics, the conclusions drawn from this study can also hold for other conventional composites reinforced with particles and short fibers. This study can be further extended to integrate additional processing and material parameters of PNCs. Finally, the present computational approach offers a technical tool for the possibility of efficient computer-aided nanocomposite design for targeted mechanical properties and quality-controllable nanocomposite manufacturing.

Acknowledgments: Partial support of the research by the NASA EPSCoR (NASA Grant #NNX07AK81A, seed grant: 43500-2490-FAR018640), NDSU Development Foundation (Grant No.: 43500-2490-FAR0017475), and NDSU Faculty Research Initial Grant is gratefully appreciated.

Author Contributions: Xiang-Fa Wu conceived and designed the computational study; Arifur Rahman conducted all the FEM simulations and figure plotting; Xiang-Fa Wu and Arifur Rahman analyzed the data; Xiang-Fa Wu wrote the paper.

Conflicts of Interest: The authors declare no conflict of interest.

References

1. Fukushima, Y.; Okada, A.; Kawasumi, M.; Kurauchi, T.; Kamigaito, O. Swelling behavior of montmorillonite by poly-6-amide. *Clay Miner.* **1998**, *23*, 27–34. [CrossRef]
2. Usuki, A.; Kojima, Y.; Kawasumi, M.; Okada, A.; Fukushima, Y.; Kurauchi, T.; Kamigaito, O. Synthesis of nylon-6-clay hybrid. *J. Mater. Res.* **1993**, *8*, 1179–1183. [CrossRef]
3. Kojima, Y.; Usuki, A.; Kawasumi, M.; Okada, A.; Kurauchi, T.; Kamigaito, O. Synthesis of nylon-6-clay hybrid by montmorillonite intercalated with ε-caprolactam. *J. Polym. Sci. Part A Polym. Chem.* **1993**, *31*, 983–986. [CrossRef]
4. Alexandre, M.; Dubois, P. Polymer-layered silicate nanocomposites: Preparation, properties and uses of a new class of materials. *Mater. Sci. Eng. R* **2000**, *28*, 1–63. [CrossRef]
5. Ray, S.S.; Okamoto, M. Polymer/layered silicate nanocomposites: A review from preparation to processing. *Prog. Polym. Sci.* **2003**, *28*, 1359–1641.
6. Tjong, S.C. Structural and mechanical properties of polymer nanocomposites. *Mater. Sci. Eng. R* **2006**, *53*, 73–197. [CrossRef]
7. Balazs, A.C.; Emrick, T.; Russell, T.P. Nanoparticle polymer composites: Where two small worlds meet. *Science* **2006**, *314*, 1107–1110. [CrossRef] [PubMed]
8. Schadler, L.S.; Brinson, L.C.; Sawyer, W.G. Polymer nanocomposites: A small part of the story. *JOM* **2007**, *59*, 53–60. [CrossRef]
9. Paul, D.R.; Robeson, L.M. Polymer nanotechnology: Nanocomposites. *Polymer* **2008**, *49*, 3187–3204. [CrossRef]
10. Chivrav, F.; Pollet, E.; Averous, K. Progress in nano-biocomposites based on polysaccharides and nanoclays. *Mater. Sci. Eng. R* **2009**, *67*, 1–17. [CrossRef]
11. Bertolino, V.; Cavallaro, G.; Lazzara, G.; Merli, M.; Milioto, S.; Parisi, F.; Sciascia, L. Effect of the biopolymer charge and the nanoclay morphology on nanocomposite materials. *Ind. Eng. Chem. Res.* **2016**, *55*, 7373–7380. [CrossRef]
12. Makaremi, M.; Pasbakhsh, P.; Cavallaro, G.; Lazzara, G.; Aw, Y.K.; Lee, S.M.; Milioto, S. Effect of morphology and size of halloysite nanotubes on functional pectin bionanocomposites for food packaging applications. *ACS Appl. Mater. Interfaces* **2017**, *9*, 17476–17488. [CrossRef] [PubMed]
13. Zabihi, O.; Ahmadi, M.; Nikafshar, S.; Preyeswary, K.C.; Naebe, M. A technical review on epoxy-clay nanocomposites: Structure, properties, and their applications in fiber reinforced composites. *Composites B* **2018**, *135*, 1–24. [CrossRef]
14. Luo, J.J.; Daniel, I.M. Characterization and modeling of mechanical behavior of polymer/clay nanocomposites. *Compos. Sci. Technol.* **2003**, *63*, 1607–1616. [CrossRef]
15. Tucker, C.L.; Liang, E. Stiffness predictions for unidirectional short-fiber composites: Review and evaluation. *Compos. Sci. Technol.* **1999**, *59*, 655–671. [CrossRef]
16. Tsai, J.; Sun, C.T. Effect of platelet dispersion on the load transfer efficiency in nanoclay composites. *J. Compos. Mater.* **2004**, *38*, 567–579. [CrossRef]
17. Weon, J.I.; Sue, H.J. Effects of clay orientation and aspect ratio on mechanical behavior of nylon-6 nanocomposite. *Polymer* **2005**, *46*, 6325–6334. [CrossRef]
18. Sheng, N.; Boyce, M.C.; Parks, D.M.; Rutledge, G.C.; Abes, J.I.; Cohen, R.E. Multiscale micromechanical modeling of polymer/clay nanocomposites and the effective clay particle. *Polymer* **2004**, *45*, 487–506. [CrossRef]

19. Zhu, L.J.; Narh, K.A. Numerical simulation of the tensile modulus of nanoclay-filled polymer composites. *J. Polym. Sci.* **2004**, *42*, 2391–2406. [CrossRef]
20. Dai, Y.; Mai, Y.W.; Ji, X. Predictions of stiffness and strength of nylon 6/MMT nanocomposites with an improved staggered model. *Compos. Part B* **2008**, *39*, 1062–1068. [CrossRef]
21. Dong, Y.; Bhattacharyya, D. A simple micromechanical approach to predict mechanical behavior of polypropylene/organoclay nanocomposites based on representative volume element (RVE). *Comput. Mater. Sci.* **2010**, *49*, 1–8. [CrossRef]
22. Hu, H.; Onyebueke, L.; Abatan, A. Characterizing and modeling mechanical properties of nanocomposites-review and evaluation. *J. Miner. Mater. Charact. Eng.* **2010**, *9*, 275–319. [CrossRef]
23. Logan, J.D. *Applied Mathematics*, 2nd ed.; Wiley: New York, NY, USA, 1997.

© 2017 by the authors. Licensee MDPI, Basel, Switzerland. This article is an open access article distributed under the terms and conditions of the Creative Commons Attribution (CC BY) license (http://creativecommons.org/licenses/by/4.0/).

Journal of
composites science

MDPI

Article

Polylactic Acid (PLA)/Cellulose Nanowhiskers (CNWs) Composite Nanofibers: Microstructural and Properties Analysis

Wenqiang Liu [1,2], Yu Dong [3], Dongyan Liu [4], Yuxia Bai [5] and Xiuzhen Lu [6,*]

1   Non-Equilibrium Metallic Materials Division, Institute of Metal Research (IMR), Chinese Academy of Sciences (CAS), Shenyang 110016, China; wqliu@imr.ac.cn
2   Shenyang Kejin Special Materials Co., Ltd., Institute of Metal Research (IMR), Chinese Academy of Sciences (CAS), Shenyang 110101, China
3   Department of Mechanical Engineering, School of Civil and Mechanical Engineering, Curtin University, Perth 6845, Australia; Y.Dong@curtin.edu.au
4   Titanium Alloys Division, Institute of Metal Research (IMR), Chinese Academy of Sciences (CAS), Shenyang 110016, China; dyliu@imr.ac.cn
5   Instruments' Center for Physical Science, University of Science & Technology of China, Hefei 230026, China; Baiyx@ustc.edu.cn
6   SMIT Center, School of Mechatronic Engineering and Automation, Shanghai University, Shanghai 200072, China
*   Correspondence: xzlu@staff.shu.edu.cn

Received: 9 January 2018; Accepted: 29 January 2018; Published: 30 January 2018

Abstract: Polylactic acid (PLA)/cellulose nanowhiskers (CNWs) composite nanofibers were successfully produced by electrospinning mixed PLA solutions with CNWs. Observation by means of transmission electron microscopy (TEM) confirms the uniform distribution of CNWs within the PLA nanofibers along the direction of the fiber axis. The spectra of composite nanofibers based on Fourier transform infrared spectroscopy (FTIR) reveal characteristic hydroxyl groups as evidenced by absorption peaks of CNWs. The addition of hydrophilic CNWs is proven to improve the water absorption ability of PLA nanofibers. The initial cold crystallization temperature decreases with the increasing CNW content, implying the nucleating agent role of CNWs as effective nanofillers. The degree of crystallinity increases from 6.0% for as-electrospun pure PLA nanofibers to 14.1% and 21.6% for PLA/5CNWs and PLA/10CNWs composite nanofibers, respectively. The incorporation of CNWs into PLA is expected to offer novel functionalities to electrospun composite nanofibers in the fields of tissue engineering and membranes.

Keywords: biopolymers and renewable polymers; crystallization; hydrophilic polymers; electrospinning; nanoparticles; nanowires and nanocrystals

---

1. Introduction

Cellulose is the most abundant natural polymer on Earth. Recently, researchers have shown a growing interest in nanocellulose research due to its numerous advantages, including biocompatibility, biodegradability, and unique chemical and reactive surface properties. Polymer composites reinforced with cellulose nanowhiskers (CNWs) have become quite attractive due to their excellent properties [1].

Electrospinning is a simple and cost-effective method for preparing polymer nanofibers. The nanofibrous mats show particular characteristics such as large surface-area-to-volume ratio and high porosity with small pore size, and have a variety of applications for filtration, sensors, electrode materials, drug delivery, cosmetics, and tissue scaffolding [2,3]. However, electrospun nanofibrous mat is not stiff enough and is sometimes difficult to handle. The incorporation of stiff nanoreinforcements, including carbon nanotubes (CNTs), nanoclays, and CNWs with high aspect ratio, has been regarded as an effective approach to enhance the mechanical, electrical, and magnetic properties of electrospun fibers [4–9].

PLA, as one of the most used biopolymers, has been widely known as an effective material candidate for electrospinning because it has good mechanical properties, biodegradability, and biocompatibility. CNWs are nanocellulosic short fibers with a length of several hundred nanometers. The incorporation of rigid CNWs into electrospun polymer nanofibers can greatly improve the mechanical properties of matrix fibers. The tensile strength and modulus of poly (ε-caprolactone) (PCL) nanofibers were increased by 68% and 37%, respectively, with the addition of 2.5 wt % CNWs [9]. The tensile strength and modulus of PLA nanofibers were improved by 5 and 22 times with the addition of 5 wt % CNWs [10]. CNWs demonstrate a good reinforcing effect to electrospun nanofibers due to their alignment within the matrix fibers. The hydrophobicity of electrospun PLA nanofibers restricts their application; the addition of hydrophilic CNWs will improve the water absorbance of PLA nanofibers, which is beneficial to their application in tissue engineering and membrane filtration fields. However, information on CNW dispersion and distribution in electrospun nanofibers has rarely been investigated. The aggregates in the center and on the surface of matrix nanofibers were observed in different studies in the literature [11,12]. In our paper, the distribution of cellulose nanowhiskers in PLA nanofibers was clearly observed under TEM (transmission electron microscope) by negatively staining with osmium tetroxide vapor and solution of uranyl acetate, respectively.

In this study, CNWs were extracted from flax yarn by using a sulfuric acid method. They were incorporated with PLA to produce composite nanofibrous mats. The nanofibrous mats were successfully prepared by electrospinning mixtures of CNWs and PLA solution. The morphological structures, thermal properties, and water absorption behavior of pure PLA and PLA/CNWs composite nanofiber mats were investigated to holistically evaluate the multifunctional properties of such composite nanofibers.

2. Experimental

2.1. Materials

Bleached flax yarns were purchased from Jayashree Textiles, Kolkata, India. Microcrystalline cellulose powders were supplied by Sigma-Aldrich Inc., St. Louis, MO, USA. Sodium hydroxide was purchased from Ajax Finechem Pty Ltd., Taren Point, Australia. Sulphuric acid with a concentration of 95–97% and chloroform (purity 99.0–99.4%, lab grade) were supplied by Merck KGaA, Darmstadt, Germany. Poly(lactic acid) (PLA, 2002 D), was purchased in pellet form from Natureworks Co., Minnetonka, MN, USA. *N,N*-dimethylformamide (DMF, Anhydrous, 99.8%) and tetrahydrofuran (THF) were obtained from Sigma-Aldrich, USA and ECP Ltd., Romil, UK, respectively.

2.2. CNWs Preparation

CNWs were isolated from flax fibers via an acid hydrolysis method. Bleached flax yarns were boiled in distilled water for 30 min and oven-dried to a constant weight. Dried yarns were then soaked for 30 min in 1% (w/w) NaOH aqueous solution at 80 °C, and washed again in running water (i.e., in an alkali-free environment). An aqueous suspension of CNWs was prepared by acid hydrolysis for 1 h in 60 wt % sulfuric acid at 55 °C with continuous stirring. After the completion of hydrolysis, the flask with the suspension was cooled in ice-cold water. The aqueous suspension of fibers was further diluted and repeatedly washed by centrifugation prior to neutralization with 1.0 wt % NaOH aqueous solution. The suspension was then freeze-dried for 48 h before use. The entire process is elaborated in Figure 1.

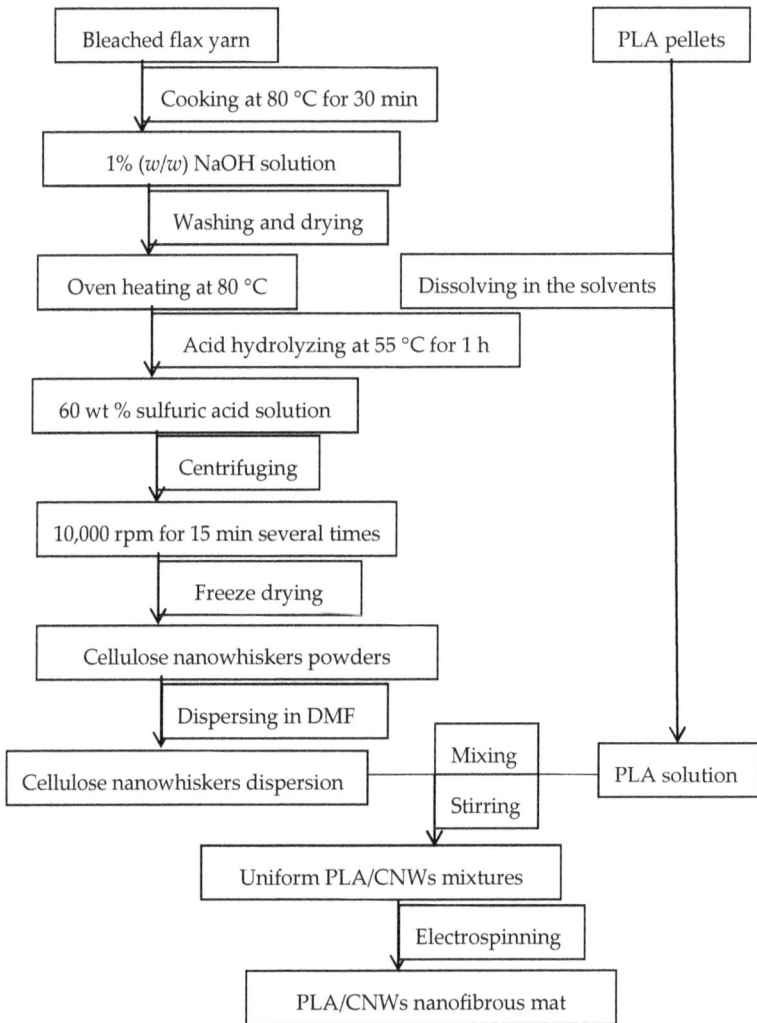

Figure 1. Flow chart for manufacturing cellulose nanowhiskers (CNWs) and electrospun polylactic acid (PLA)/CNWs composite nanofibrous mat.

The concentrated aqueous CNWs suspension and freeze-dried CNWs powders are depicted in Figure 2. The incorporation of sulphate groups along the surface of the crystallites results in a negative charge of the surface. This anionic stabilization via attraction/repulsion forces of electrical double layers at the crystallites is the main reason for the stability of colloidal suspensions of crystallites. The loosely packed powders are easily dispersed in water or solvents with ultrasonic treatment, which is feasible for obtaining homogeneous mixtures with polymer solutions.

Figure 2. Aqueous CNWs suspension and freeze-dried CNWs powders.

2.3. Preparation of PLA/CNWs Nanofibrous Mat

A weighed amount of CNWs was dispersed in DMF at ambient temperature. Then, appropriate amounts of THF and PLA were added and the mixture was stirred for several hours. The final dispersion contained 5.0 or 10.0 wt % CNWs with respect to the total amount of PLA and CNWs, and 12% PLA/CNWs solution. The masses of components and solvents are presented in Table 1. Electrospun mats were obtained by applying a high voltage between a needle tip and a grounded collector. Nanofibrous mats were collected on an aluminum foil by a needle tip-to-collector distance of 100 mm with an applied voltage of 10–15 kV and a solution feed rate of 1.0–1.5 mL h^{-1}. When a high voltage was applied to the metal syringe needle, electrical charge was built up on the solution surface. The charge was attracted towards an electrically grounded collector, in our case, covered by a piece of aluminum foil. As the charge travelled to the grounded collector, an electrified thin jet of polymer solution was pulled from the needle. After the solution left the syringe, the solvent evaporated rapidly and a very thin stream of elongated polymer fibers went toward the collector. Thin fibers were thereby randomly deposited on the collector surface. The electrospun nanofibrous mat was peeled with care from the aluminum foil for subsequent testing use.

Table 1. Mass concentration of electrospun PLA and PLA/CNWs composites.

CNWs (wt %)	CNWs (g)	PLA (g)	DMF (g)	THF (g)
0	0	1.0	5.5	1.8
5	0.05	0.95	5.5	1.8
10	0.1	0.90	5.5	1.8

2.4. Characterization Methods

The structures of CNWs and PLA/CNWs nanofibers were investigated using a transmission electron microscope (TEM—Philips CM 12, Holland, The Netherlands) at the accelerating voltage of 100 kV. A droplet of the diluted suspension was allowed to float on and eventually flow through a copper grid covered with a carbon film. The samples were then stained by allowing the grids to float in a 2.0 wt % solution of uranyl acetate for 1 min. TEM observation of composite nanofibers was performed by mounting the fibrous mat in epoxy resin, followed by negatively staining with osmium tetroxide vapor and solution of uranyl acetate, respectively.

X-ray diffraction (XRD–Bruker D 8 Advance, Karlsruhe, Germany) measurements were performed using Cu Kα radiation (scan speed of $0.02° \cdot s^{-1}$ from the diffraction angle $2\theta = 5°–50°$) at 40 kV and 40 mA.

The thermal behavior of the fibers was characterized by a differential scanning calorimeter (DSC, TA instrument Q1000, New Castle, DE, USA) using a heat/cooling cycle between 0 and 180 °C at a heating/cooling rate of 10 °C/min. A Fourier transform infrared (FTIR Nicolet 8700, Mettler-Toledo, LLC, Columbus, OH, USA) spectroscope based on attenuated total reflectance (ATR) was employed to analyze the chemical structures of the samples. All spectra were collected with 4 cm^{-1} wave number resolution after 64 continuous scans at a wavelength range of 4000–600 cm^{-1}.

Thermal stability was assessed with a thermogravimetric analysis (TGA, Q5000, New Castle, DE, USA). Samples were heated in open platinum pans from room temperature to 600 °C, under a nitrogen atmosphere to avoid thermoxidative degradation, at a heating rate of 10 °C/min.

The degree of crystallinity (X) was evaluated from the DSC data according to Equation (1) [13]:

$$x\% = \frac{\Delta H_m - \Delta H_c}{\Delta H_m^0 \times X_{PLA}} \times 100\% \tag{1}$$

where ΔH_m and ΔH_c are melting and crystallization enthalpies, respectively; ΔH_m^0 is 93.6 J/g for 100% crystalline PLA crystals [13]; and X_{PLA} is the weight fraction of PLA matrices in their composites.

The morphology of the electrospun nanofibers was studied using a field emission scanning electron microscope (FE-SEM, FEI XL30s) with an accelerating voltage of 5 kV.

To calculate the water absorption, the samples were immersed in distilled water at room temperature for 24 h after being dried in the oven at 80 °C until no weight change was observed. Subsequently, the samples were taken out, blotted with filter paper to absorb the excess water on the surface, and weighed in a precise balance. The water absorption ratio (g/g) was determined according to the following equation:

$$\text{Water absorption ratio} = (W_t - W_0)/W_0 \tag{2}$$

where W_0 is the mass of the dried sample and W_t is the mass of the swollen sample at time t (here t = 24 h).

3. Results and Discussion

3.1. Morphological Characterizations of CNWs

The average diameter and length for individual cellulose fibers were measured to be 20 and 300 nm, respectively. The sizes of CNWs are dependent on the source of raw materials and their production methods. CNWs from cotton show lower aspect ratios between 10–12 [14,15], as opposed to that of CNWs from tunicin at 200 [16]. TEMPO-oxidization can produce CNWs with a high aspect ratio relative to those produced using sulfuric acid hydrolysis [17]. As it is well known, fillers with high aspect ratios conventionally are crucial for an efficient reinforcement effect in polymer matrices.

3.2. Characterization of Electrospun PLA/CNWs Composite Nanofibers

3.2.1. Morphology

CNWs show a significant reinforcing effect in polymeric nanofibers, which results from the synergistic effect of solution flow and electric charge that induced the orientation of nanofillers along the fiber tailored direction during the electrospinning process. The reinforcement of fillers is strongly dependent on good filler distribution within matrix fibers. The fibrous morphology is usually observed via scanning electron microscopy (SEM). The SEM morphology of pure PLA and PLA/CNWs composite nanofibrous mats produced by electrospinning is depicted in Figure 3, and is clearly indicative of typical porous structures. Pure PLA nanofibers are relatively uniform with average fiber diameters of approximately 202 nm. The diameters of PLA/10CNWs composite nanofibers are slightly smaller (at 129 nm) than those of PLA/5CNWs counterparts (at 170 nm) due to the decrease in the total concentration of mixtures, resulting from the replacement of PLA with CNWs.

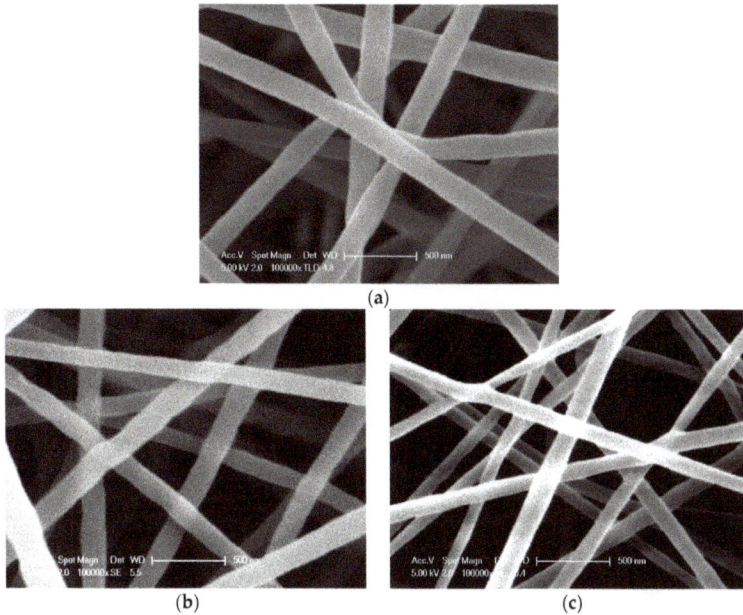

Figure 3. SEM micrographs of electrospun nanofibers: (**a**) pure PLA; (**b**) PLA/5CNWs composites; and (**c**) PLA/10CNWs composites.

No CNWs appear to protrude from the outer surface of the PLA fibers, implying that they are completely embedded into the PLA matrices. They may be aligned along the fiber axis through the entire fibers during the electrospinning process because CNWs can also be aligned in the electrical field [18]. Similar behavior has been observed for electrospun polymer/CNT composite nanofibers [19,20]. This distribution behavior makes a significant contribution to the mechanical strength and modulus of matrix fibers. It arises from more effective reinforcements of nanofillers within matrices as well as better interfacial bonding between nanofillers and matrix fibers, when compared with polymer nanocomposites produced by solution casting and melt compounding methods.

The existence of CNWs in morphological structures is further detected under TEM, as exhibited in Figure 4a. The sulfuric acid hydrolysis of cellulose causes the breakdown of the fibers into rod-like fragments with a diameter of 20 nm and length of 300 nm. Amorphous phases were selectively hydrolyzed, and crystalline phases remained unaffected. Stained CNWs (i.e., dark dots represent the cross sections of CNWs) are well aligned along PLA nanofibers (i.e., long bright fibers) and uniformly distributed within matrix fibers. The distribution of nanofillers in electrospun nanofibers is associated with the material selection of solvents and polymer matrices. CNWs can be uniformly dispersed in polar-group solvents, such as water, DMF, etc. [21,22]. Rojas et al. found that 9 wt % CNWs were distributed on the outer surface of polystyrene (PS) nanofibers due to centrifugal effects during electrospinning [11]. CNWs aggregated in the center of polymethyl methacrylate (PMMA) nanofibers by using an etching method as mentioned by Dong et al. [12].

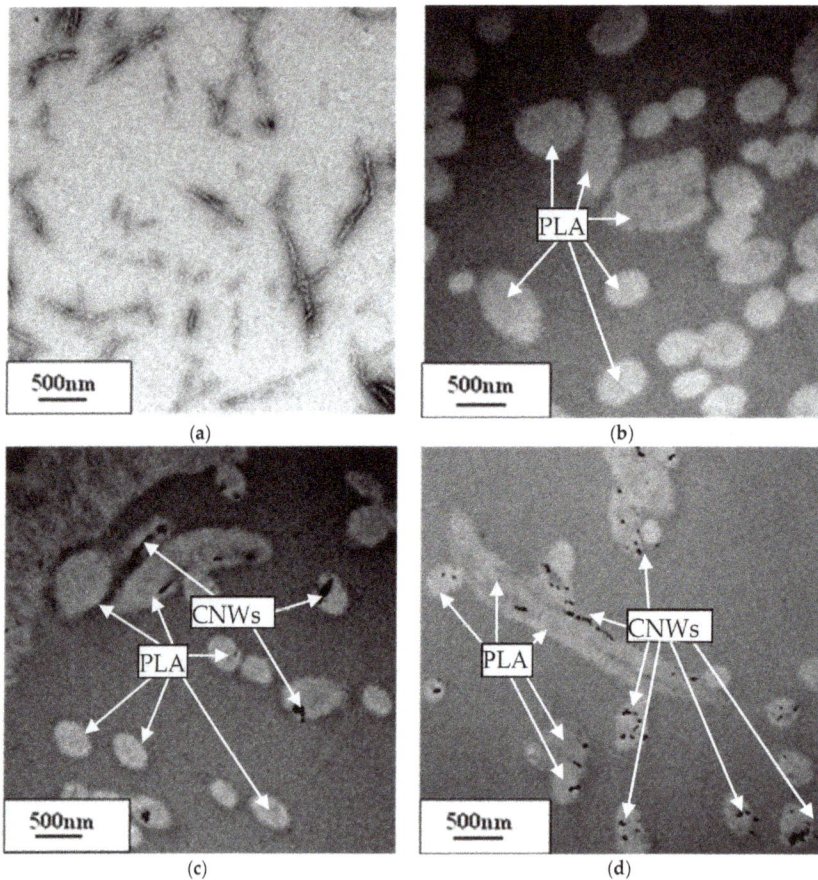

Figure 4. TEM micrographs of morphological structures: (**a**) CNWs; (**b**) pure PLA; (**c**) PLA/5CNWs composites; and (**d**) PLA/10CNWs composites.

3.2.2. Structural Analysis

FTIR analysis was carried out to confirm the presence of CNWs in PLA nanofiber mats and further assess their interactions. PLA is a semicrystalline polymer with the chemical formula of $(C_3H_4O_2)_n$. As shown in Figure 5, The needle-like peaks at 1737 and 1165 cm^{-1} are assigned to the carbonyl stretching C=O and stretching vibration of C–O in PLA chains. A mountainous triplet of peaks at 1110, 1072, and 1025 cm^{-1} correspond to C–O stretching vibrations [23]. The absorption bands at 921 and 908 cm^{-1} are characteristic of PLA/CNWs composite nanofibers. This phenomenon signifies their prevalent amorphous structures [24]. However, according to the relative decrease in the peak intensity of the amorphous phase at 955 cm^{-1}, it is elucidated that the degree of crystallinity increases as the CNW content increases. Weak bands between 3550 and 3200 cm^{-1}, which are assigned to typical stretching –OH vibrations in cellulose, are present in the spectra of composite mats with the addition of 5.0 and 10.0 wt % CNWs. This finding confirms the incorporation of CNWs into PLA matrix fibers during the electrospinning process. The peaks of composite nanofibers show no difference from those of pure PLA, which suggests that little interaction takes place between CNWs and PLA.

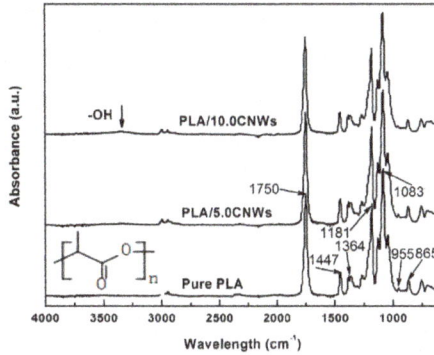

Figure 5. FTIR spectra of PLA/CNWs composite nanofibers.

Figure 6 shows XRD patterns of pure PLA and PLA/CNWs composite mats. Pure PLA mat exhibits a broad diffraction pattern without the obviously sharp peaks of crystalline PLA, indicating their dominant amorphous structures. This means that no detectible crystallization for the pure PLA nanofibers occurs due to the rapid evaporation rate of solvents. Moreover, polymeric chains under the high elongation rate have less time to form crystalline lamellae, leading to a lower degree of crystallinity. However, sharp peaks appear for composite mats with the inclusion of 5.0 and 10.0 wt % CNWs. The tiny peak presented at 22.4° is the typical diffraction of crystalline CNWs because CNWs generally possess two typical peaks at 15°–16° and 22.4°, and the latter is more intense [25]. The intense peak at 16.6° for PLA/10CNWs is evident, which belongs to the peaks of crystalline PLA to enhance the crystallinity degree of PLA [26]. This peak is much more pronounced than that of PLA nanofibers with addition of 7.5% CNWs [27]. The increased crystallinity degree of PLA is attributed to the inclusion of a high content of CNWs. The degree of crystallinity of PLA can be calculated according to the melting enthalpy in further-mentioned DSC curves.

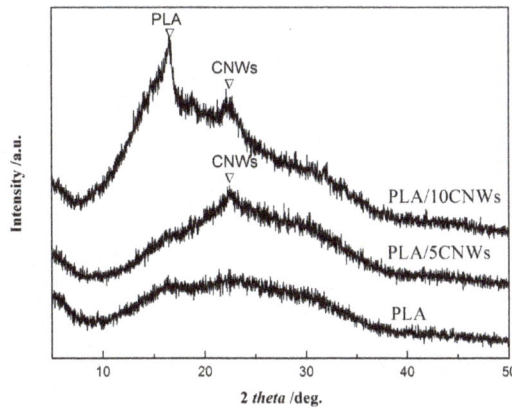

Figure 6. XRD patterns of pure PLA nanofibers and PLA/CNWs composite nanofibers.

3.2.3. DSC Characterizations

Thermal properties of electrospun nanofibers can be determined from the characteristic exothermic or endothermic reactions on DSC curves. When heated, electrospun fibers exhibit glass transition, followed by cold crystallization and melting processes. The thermal behaviors of electrospun nanofibers shown in Figure 5 are quite similar to those of PLA/CNWs composite nanofibers [28]. The glass transition temperatures (T_g) and melting temperatures (T_m) of all electrospun nanofibers occur at about 55 and 150 °C, respectively, regardless of processing methods and CNWs concentration. However, their characteristic crystallization parameters demonstrate a great difference from bulk films prepared by solution casting and melting compounding methods. The onset crystallization temperatures (T_c) were found to be around 69.7, 68.2, and 68.8 °C, for neat PLA nanofibers as well as PLA/5CNWs and PLA /10CNWs composite nanofibers, respectively. These temperatures for as-spun nanofibers are over 30 °C lower than those composite films manufactured using solution cast and melt compounding (over 100 °C) [28,29]. This is because molecular chains in electrospun PLA nanofibers are highly oriented compared to the randomly coiled chains in PLA cast films [30]. The rapid drawing and solidification of polymer jets leads to the nonequilibrium conformation and highly orientated polymer chains along the axis of the long matrix nanofibers during the electrospinning process. Such imperfection gives rise to decreases in crystallization temperatures. Oh et al. reported that the onset temperature of cold crystallization decreases as the draw ratio increases, resulting from the initial cold crystallization behavior of PLA matrices that is enhanced by the presence of strain-induced crystallinity [31,32]. The crystallization takes place at lower temperatures immediately after the glass transition, which can be attributed to the presence of numerous crystal nuclei. Therefore, electrospun nanofibers showed different properties from those of PLA/CNWs composite films with a low degree of crystallinity prepared by melt compounding. The overlapping crystallization peaks with a wide temperature range of electrospun nanofibers also vary from those of films obtained via conventional solution casting and melt compounding with a single exothermic peak. This phenomenon indicates the complexity of molecular structures of electrospun nanofibers. It may be associated with the formation of polymorphic crystals, which will be further investigated in our subsequent work.

The effects of CNWs on the cold crystallization of PLA nanofibers and PLA/CNWs composite nanofibers are shown in Figure 7a. The cold crystallization of neat PLA mat appears to present two overlapping peaks. The decrease in T_c values with the addition of CNWs may result from the existence of more nuclei because CNWs can be effective nucleating agents to accelerate the PLA cold crystallization process [33,34]. The degree of crystallinity in electrospun PLA mat is very low as listed in Table 1, indicating that the majority of the chains are in the amorphous phase. It can be easily understood that the rapid solidification of stretched chains under a high elongation rate at the later stage of electrospinning may hinder the crystal development owing to insufficient time for molecular chains to form crystal structures. These two crystallization peaks were observed to shift to lower temperatures at 69.7 and 93.4, 68.2 and 92.2, as well as 68.8 and 91.4 °C for pure PLA nanofibers and PLA/CNWs composite nanofibers at CNWs contents of 5.0 and 10 wt %, respectively, as shown in Table 2.

Table 2. Thermal parameters of PLA nanofibers and PLA/CNWs composite nanofibers.

CNW Content (wt %)	T_g (°C)		T_c (°C)		ΔH_c (J/g^{-1})		ΔH_m (J/g^{-1})		X (%)	
	1st	2nd	1st	2nd	1st	2nd	1st	2nd	1st	2nd
0	55.6	56.4	69.7	93.4	21.0	-	26.6	0.19	6.0	0.2
5.0	55.4	56.9	68.2	92.2	14.5	1.37	27.0	1.9	14.1	0.6
10.0	56.1	57.2	68.8	91.0	4.5	7.46	22.7	10.4	21.6	3.5

In order to eliminate the thermal history for the samples, the fibers were heated above the melting temperature of 180 °C during the first heating cycle, and cooled below glass temperature and reheated above the T_g again during the second heating cycle. The DSC curves are shown in Figure 7b. The curves at the second heating cycle demonstrate significant differences in cold crystallization and melting peaks from those at the first cycle. Apparently, as-spun nanofibers with well-oriented molecular chains can be easily destroyed by transforming fibrous structures into polymeric melt upon the melting stage. Pure PLA nanofibers were shown to undergo multiple stages of thermal effect from glass transition, weak cold crystallization to a melting process with a weak and wide melting peak observed at 151.2 °C, whereas PLA/CNWs composite nanofibers reveal distinct high T_c values of 113.8 and 111.3 °C, as well as T_m values of 150.5 and 149.2 °C for PLA/CNWs composite nanofibers reinforced with 5% and 10% CNWs, respectively. The apparent cold crystallization of composite mats is due to embedded CNWs facilitating the nucleation of PLA crystals in the cold crystallization process, thus leading to the decrease of T_c and the increase in the degree of crystallization. This nucleating effect of CNWs manifested in both electrospun nanofibers and subsequently quenched samples in that composite mat induce increases in the degree of crystallinity in both heating cycles. The degree of crystallinity increases from 6.0% for electrospun PLA nanofibers to 14.1% and 21.6% (tabulated in Table 2) for PLA/5CNWs and PLA/10CNWs composite nanofibers, respectively. These crystallinity values are consistent with the XRD results. On the other hand, the degrees of crystallinity for the quenched samples at the second heating cycle are only 0.2%, 0.6%, and 3.5% for pure PLA nanofibers, PLA/5CNWs, and PLA/10CNWs composite nanofibers, respectively, which are much lower than those of the counterparts at the first heating cycle. The orientation of molecular chains and nanofiber elongation during electrospinning are responsible for the increase in the degree of crystallinity for electrospun nanofibers.

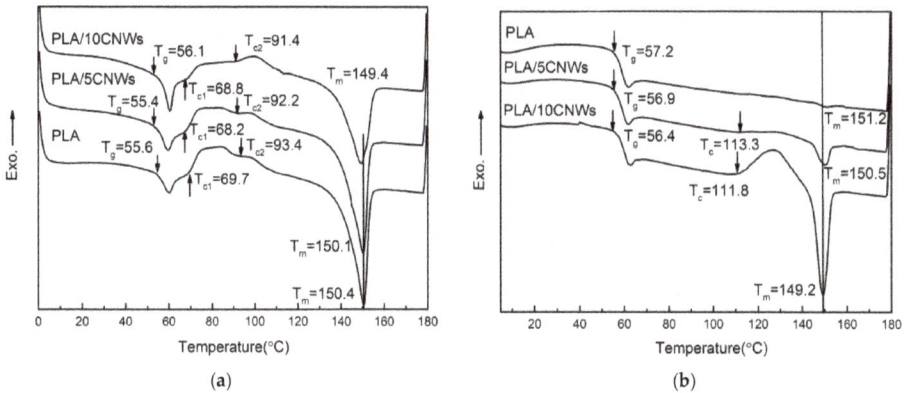

Figure 7. DSC curves of PLA and PLA/CNWs composite mats: (**a**) first heating cycle and (**b**) second heating cycle.

3.2.4. Thermal Degradation

Figure 8 shows TGA and DTGA (derivative TGA) curves of PLA and PLA/CNWs composite mats. The main degradation behaviors of PLA and PLA/CNWs composite mats are similar. The decomposition starts at around 260 °C, followed by a rapid weight loss of over 96% at the temperatures around 350–360 °C. The initial decomposition temperature is dependent on the molecular weight and crystallinity of PLA. A slightly higher weight remains for composite mats due to the existence of CNWs, which are more stable at the temperature range over 310 °C than is pure PLA. The composite nanofibers including cellulose nanowhiskers show a little more weight loss than neat PLA nanofibers at the temperature below 300 °C, which is ascribed to the faster degradation rate of cellulose nanowhiskers than of PLA nanofibers. The weight loss increases with an increase in the concentration of cellulose nanowhiskers. Above 300 °C, the composite nanofibers show improved thermal stability due to the low degradation rate of cellulose nanowhiskers, corresponding to the lower peak intensity and the shift peak position to a high temperature in the DTGA curves [35].

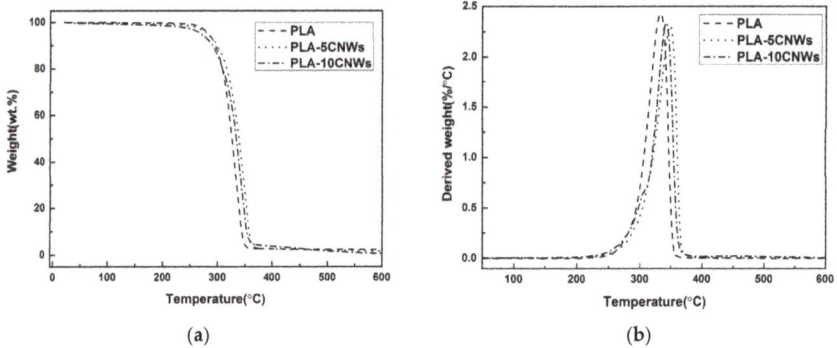

Figure 8. PLA and PLA/CNWs composite mats: (**a**) TGA curves and (**b**) DTGA (derivative TGA) curves.

3.2.5. Water Absorption

In addition to the reinforcing effect of nanofillers, the water wettability of polymer matrices can be improved with the addition of hydrophilic CNWs. PLA has been approved by Food and Drug Administration (FDA) for clinical use such as in surgical sutures [36]. However, the hydrophobic nature of electrospun PLA nanofibers restricts their applications in tissue scaffolding because cells may attach and proliferate less well than on matrices with good wettability [37]. The use of hydrophilic nanomaterials or polymers will improve the water wettability of hydrophobic polymeric nanofibers [38,39]. Water absorption or retention is a simple and direct gravimetric test for determining the maximum amount of fluid absorption and fluid retention on tested materials. The water absorption ratio of composite nanofibers greatly increases with an increase in the CNWs content from 2.5 to 7.5 wt %, as illustrated in Figure 6. The water absorption ratio of pure PLA nanofibrous mat is over 200% owing to its porous structure. This ratio reaches over 1600% with the addition of 7.5% CNWs, which is 8 times that of the neat PLA mat. This improvement in water absorption is beneficial to applications in tissue engineering or membranes. The immersion state of the pure PLA mat and PLA/CNWs composite mats are also shown in Figure 9. It is evidently seen that the PLA/CNWs composite mat with 5.0 and 7.5 wt % CNWs inclusions dropped down to the bottom of the vials, whereas mats of pure PLA and PLA/CNWs composites with 2.5 wt % CNWs still floated on the top surface of the water. This indirectly proves that the water absorption ability of the PLA mat is improved with the incorporation of CNWs.

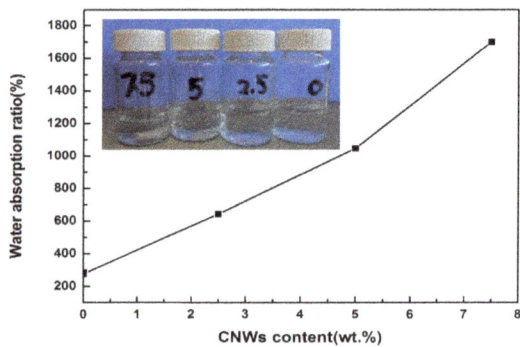

Figure 9. Variations of water absorption for electrospun PLA/CNWs composite nanofibers with different CNWs content. The inserted picture shows the mats' different floating conditions in distilled water after being immersed for 24 h.

4. Conclusions

Rod-shaped CNWs with lateral and longitudinal dimensions of 30 and 300 nm, respectively, are favorable for preparing PLA/CNWs composite nanofibers without clogging the needle during the electrospinning process. The composite nanofibers are thinner than their pure PLA counterparts due to their lower solution concentration by adding CNWs. The nanofibers show smooth surfaces without any obvious protrusion of CNWs on the outer surfaces of the PLA nanofibers. The CNWs are well aligned along the PLA nanofibers and uniformly distributed within matrix fibers. Electrospun PLA nanofibers and PLA/CNWs composite nanofibers show lower crystallization temperatures and higher degrees of crystallinity than the quenched samples. The addition of CNWs has also been found to effectively enhance the water absorption of PLA nanofibers. CNWs and their polymer composite nanofibers can offer great potential for widespread applications from biomedical engineering, to sensors, to nanofiltration.

Acknowledgments: The authors would like to thank the Key Laboratory of Advanced Display and System Applications of Ministry of Education, Shanghai University (China) for various support.

Author Contributions: W.L. and X.L. conceived and designed the experiments; W.L. and Y.B. performed the experiments; Y.D. and D.L. analyzed the data; Z.L. and Y.B. contributed reagents/materials/analysis tools; W.L., D.L., and Y.D. wrote the paper.

Conflicts of Interest: The authors declare no conflicts of interest.

References

1. Eichhorn, S.J.; Dufresne, A.; Aranguren, M.; Marcovich, N.E.; Capadona, J.R.; Rowan, S.J.; Weder, C.; Thielemans, W.; Roman, M.; Renneckar, S.; et al. Current international research into cellulose nanofibres and nanocomposites. *J. Mater. Sci.* **2010**, *45*, 1–33. [CrossRef]

2. Ding, B.; Wang, M.; Yu, J.; Sun, G. Gas sensors based on electrospun nanofibers. *Sensors* **2009**, *9*, 1609–1624. [CrossRef] [PubMed]

3. Bhattaraia, S.R.; Bhattarai, N.; Yi, H.K.; Hwang, P.H.; Cha, D.I.; Kim, H.Y. Novel biodegradable electrospun membrane: Scaffold for tissue engineering. *Biomaterials* **2004**, *25*, 2595–2602. [CrossRef]

4. Hou, H.; Ge, J.J.; Zeng, J.; Li, Q.; Reneker, D.H.; Greiner, A. Electrospun polyacrylonitrile nanofibers containing a high concentration of well-aligned multiwall carbon nanotubes. *Chem. Mater.* **2005**, *17*, 967–973. [CrossRef]

5. Ji, J.; Sui, G.; Yu, Y.; Liu, Y.; Lin, Y.; Du, Z. Significant improvement of mechanical properties observed in highly aligned carbon-nanotube-reinforced nanofibers. *J. Phys. Chem. C* **2009**, *113*, 4779–4785. [CrossRef]

6. Park, W.I.; Kang, M.; Kim, H.S.; Jin, H.J. Electrospinning of poly (ethylene oxide) with bacterial cellulose whiskers. *Macromol. Symp.* **2007**, *249–250*, 289–294. [CrossRef]

7. Peresin, M.S.; Habibi, Y.; Zoppe, J.O.; Pawlak, J.; Rojas, O.J. Nanofiber composites of polyvinyl alcohol and cellulose nanocrystals: Manufacture and characterization. *Biomacromolecules* **2010**, *11*, 674–681. [CrossRef] [PubMed]

8. Hong, J.H.; Jeong, E.H.; Lee, H.S.; Baik, D.H.; Seo, S.W.; Youk, J.H. Electrospinning of polyurethane/organically modified montmorillonite nanocomposites. *J. Polym. Sci. Part B Polym. Phys.* **2005**, *43*, 3171–3177. [CrossRef]

9. Zoppe, J.O.; Peresin, M.S.; Habibi, Y.; Venditti, R.; Rojas, O.J. Reinforcing poly (ε-caprolactone) nanofibers with cellulose nanocrystals. *Appl. Mater. Interfaces* **2009**, *1*, 1996–2004. [CrossRef] [PubMed]

10. Shi, Q.F.; Zhou, C.J.; Yue, Y.Y.; Guo, W.; Wu, Y.; Wu, Q. Mechanical properties and in vitro degradation of electrospun bio-nanocomposite mats from PLA and cellulose nanocrystals. *Carbohyd. Polym.* **2012**, *90*, 301–308. [CrossRef] [PubMed]

11. Rojas, O.J.; Montero, G.A.; Habibi, Y. Electrospun nanocomposites from polystyrene loaded with cellulose nanowhiskers. *J. Appl. Polym. Sci.* **2009**, *113*, 927–935. [CrossRef]

12. Dong, H.; Strawhecker, K.E.; Snyder, J.F.; Orlicki, J.A.; Reiner, R.S.; Rudie, A.W. Cellulose nanocrystals as a reinforcing material for electrospun poly(methyl methacrylate) fibers: Formation, properties and nanomechanical characterization. *Carbohd. Polym.* **2012**, *87*, 2488–2495. [CrossRef]

13. Ke, T.Y.; Sun, X.Z. Physical properties of poly (lactic acid) and starch composites with various blending ratios. *Cereal. Chem.* **2000**, *77*, 761–768. [CrossRef]
14. Martins, M.; Teixeira, E.M.; Corrêa, A.C.; Ferreira, M.; Mattoso, L.H.C. Extraction and characterization of cellulose whiskers from commercial cotton fibers. *J. Mater. Sci.* **2011**, *46*, 7858–7864. [CrossRef]
15. Roohani, M.; Habibi, Y.; Belgacem, N.; Ebrahim, G.; Karimi, A.; Dufresne, A. Cellulose whiskers reinforced polyvinyl alcohol copolymers nanocomposites. *Eur. Polym. J.* **2008**, *44*, 2489–2498. [CrossRef]
16. Angles, M.N.; Dufresne, A. Plasticized starch/tunicin whiskers nanocomposites. 1. Structural analysis. *Macromolecules* **2000**, *33*, 8344–8353. [CrossRef]
17. Sun, X.; Wu, Q.; Ren, S.; Lei, T. Comparison of highly transparent all-cellulose nanopaper prepared using sulfuric acid and TEMPO-mediated oxidation methods. *Cellulose* **2015**, *22*, 1123–1133. [CrossRef]
18. Habibi, Y.; Heim, T.; Douillard, R. AC electric field-assisted assembly and alignment of cellulose nanocrystals. *J. Polym. Sci. Part B Polym. Phys.* **2008**, *46*, 1430–1436. [CrossRef]
19. Lu, P.; Hsieh, Y.L. Multiwalled carbon nanotube (MWCNT) reinforced cellulose fibers by electrospinning. *Appl. Mater. Interfaces* **2010**, *2*, 2413–2420. [CrossRef] [PubMed]
20. Kannan, P.; Eichhorn, S.J.; Young, R.J. Deformation of isolated single-wall carbon nanotubes in electrospun polymer nanofibers. *Nanotechnol* **2007**, *18*, 235707. [CrossRef]
21. Beck, S.; Bouchard, J.; Berry, R. Dispersibility in water of dried nanocrystalline cellulose. *Biomacromolecules* **2012**, *13*, 1486–1494. [CrossRef] [PubMed]
22. Van den Berg, O.; Capadona, J.R.; Weder, C. Preparation of homogeneous dispersions of tunicate cellulose whiskers in organic solvents. *Biomacromolecules* **2007**, *8*, 1353–1357. [CrossRef] [PubMed]
23. Mai, T.T.T.; Nguye, T.T.T.; Le, Q.D.; Nguyen, T.N.; Ba, T.C.; Nguyen, H.B.; Phan, T.B.H.; Tran, D.L.; Nguyen, X.P.; Park, J.S. A novel nanofiber Cur-loaded polylactic acid constructed by electrospinning. *Adv. Nat. Sci. Nanosci. Nanotechnol.* **2012**, *3*, 025014.
24. Ribeiro, C.; Sencadas, V.; Costa, C.M.; Ribelles, J.L.G.; Lanceros-Mendez, S. Tailoring the morphology and crystallinity of poly (L-lactide acid) electrospun membranes. *Sci. Technol. Adv. Mater.* **2011**, *12*, 015001. [CrossRef] [PubMed]
25. Lamaminga, J.; Hashima, R.; Sulaimana, O.; Leha, C.P.; Sugimoto, T.; Nordina, N.A. Cellulose nanocrystals isolated from oil palm trunk. *Carbohyd. Polym.* **2015**, *127*, 202–208. [CrossRef] [PubMed]
26. Yasuniwa, M.; Sakamo, K.; Ono, Y.; Kawahara, W. Melting behavior of poly (L-lactic acid): X-ray and DSC analyses of the melting process. *Polymer* **2008**, *49*, 1943–1951. [CrossRef]
27. Liu, D.Y.; Yuan, X.W.; Bhattacharyya, D. The effects of cellulose nanowhiskers on electrospun poly (lactic acid) nanofibers. *J. Mater. Sci.* **2012**, *47*, 3159–3165. [CrossRef]
28. Espino-Pérez, E.; Bras, J.; Ducruet, V.; Guinault, A.; Dufresne, A.; Domenek, S. Influence of chemical surface modification of cellulose nanowhiskers on thermal, mechanical, and barrier properties of poly (lactide) based bionanocomposites. *Eur. Polym. J.* **2013**, *49*, 3144–3154. [CrossRef]
29. Raquez, S.J.M.; Murena, Y.; Goffin, A.L.; Habibi, Y.; Ruelle, B.; DeBuyl, F.; Dubois, P. Surface-modification of cellulose nanowhiskers and their use as nanoreinforcers into polylactide: A sustainably-integrated approach. *Comp. Sci. Technol.* **2012**, *72*, 544–549. [CrossRef]
30. Zong, X.; Kim, K.; Fang, D.; Ran, S.; Hsiao, B.S.; Chu, B. Structure and process relationship of electrospun bioabsorbable nanofiber membranes. *Polymer* **2002**, *43*, 4403–4412. [CrossRef]
31. Oh, M.O.; Kim, S.H. Conformational development of polylactide films induced by uniaxial drawing. *Polym. Int.* **2014**, *63*, 1247–1253. [CrossRef]
32. Stoclet, G.; Seguela, R.; Lefebvre, J.M.; Elkoun, S.; Vanmansart, C. Strain-induced molecular ordering in polylactide upon uniaxial stretching. *Macromolecules* **2010**, *43*, 1488–1498. [CrossRef]
33. Salmeron, S.M.; Mathot, V.B.F.; Vanden, P.G.; Gomez, R.J.L. Effect of the cooling rate on the nucleation kinetics of poly(L-lactic acid) and its influence on morphology. *Macromolecules* **2007**, *40*, 7989–7997. [CrossRef]
34. Pei, A.; Zhou, Q.; Berglund, L.A. Functionalized cellulose nanocrystals as biobased nucleation agents in poly(L-lactide)(PLLA)–crystallization and mechanical property effects. *Comp. Sci. Technol.* **2010**, *70*, 815–821. [CrossRef]
35. El-Sakhawy, M.; Hassan, M.L. Physical and mechanical properties of microcrystalline cellulose prepared from agricultural residues. *Carbohyd. Polym.* **2007**, *67*, 1–10. [CrossRef]
36. Yang, S.; Leong, K.F.; Du, Z.; Chua, C.K. The design of scaffolds for use in tissue engineering. Part I. Traditional factors. *Tissue Eng.* **2001**, *7*, 679–689. [CrossRef] [PubMed]

J. Compos. Sci. **2018**, *2*, 4

37. Ma, M.; Gupta, M.; Li, Z.; Zhai, L.; Gleasom, L.L.; Cohen, R.E. Decorated electrospun fibers exhibiting superhydrophobicity. *Adv. Mater.* **2007**, *19*, 255–259. [CrossRef]
38. Cui, W.; Cheng, L.; Li, H.; Zhou, Y.; Zhang, Y.; Chang, J. Preparation of hydrophilic poly(L-lactide) electrospun fibrous scaffolds modified with chitosan for enhanced cell biocompatibility. *Polymer* **2012**, *53*, 2298–2305. [CrossRef]
39. Zhang, P.; Tian, R.; Lv, T.; Na, B.; Liu, Q. Water-permeable polylactide blend membranes for hydrophilicity-based separation. *Chem. Eng. J.* **2015**, *269*, 180–185. [CrossRef]

© 2018 by the authors. Licensee MDPI, Basel, Switzerland. This article is an open access article distributed under the terms and conditions of the Creative Commons Attribution (CC BY) license (http://creativecommons.org/licenses/by/4.0/).

Journal of
composites science

MDPI

Article

Milling of Nanoparticles Reinforced Al-Based Metal Matrix Composites

Alokesh Pramanik [1,*], Animesh Kumar Basak [2], Yu Dong [1], Subramaniam Shankar [3] and Guy Littlefair [4]

1   School of Civil and Mechanical Engineering, Curtin University, GPO Box U1987, Bentley, WA 6845, Australia; Y.Dong@curtin.edu.au
2   Adelaide Microscopy, The University of Adelaide, Adelaide, SA 5005, Australia; animesh.basak@adelaide.edu.au
3   Department of Mechatronics Engineering, Kongu Engineering College, Perundurai 638060, India; shankariitm@gmail.com
4   Faculty of Design and Creative Technologies, Auckland University of Technology, Auckland 1010, New Zealand; guy.littlefair@aut.ac.nz
*   Correspondence: alokesh.pramanik@curtin.edu.au; Tel.: +61-8-9266-7981

Received: 27 January 2018; Accepted: 1 March 2018; Published: 2 March 2018

Abstract: This study investigated the face milling of nanoparticles reinforced Al-based metal matrix composites (nano-MMCs) using a single insert milling tool. The effects of feed and speed on machined surfaces in terms of surface roughness, surface profile, surface appearance, chip surface, chip ratio, machining forces, and force signals were analyzed. It was found that surface roughness of machined surfaces increased with the increase of feed up to the speed of 60 mm/min. However, at the higher speed (100–140 mm/min), the variation of surface roughness was minor with the increase of feed. The machined surfaces contained the marks of cutting tools, lobes of material flow in layers, pits and craters. The chip ratio increased with the increase of feed at all speeds. The top chip surfaces were full of wrinkles in all cases, though the bottom surfaces carried the evidence of friction, adhesion, and deformed material layers. The effect of feed on machining forces was evident at all speeds. The machining speed was found not to affect machining forces noticeably at a lower feed, but those decreased with the increase of speed for the high feed scenario.

Keywords: Nano-MMCs; 6061 aluminum alloys; machinability; milling

---

1. Introduction

Composite materials are increasingly used in different structural applications for their high performance in services. Among different composite materials, metal matrix composites (MMCs), particularly aluminum-based particle/fiber-reinforced composites have a high strength to weight ratio and wear resistance, and thus are increasingly used in automotive and aerospace structures [1,2]. Metal matrix composites (MMCs) are generally reinforced with micro-sized ceramic/oxide reinforcements in the shape of fibers or particles [3]. The sizes of reinforcing particulates in MMCs range from a few to several hundred micrometers [4]. The research on discontinuous particulate/fiber-reinforced MMCs has been a focus because of their low manufacturing cost, ease of production, and macroscopically isotropic mechanical properties [5,6]. Particulate-reinforced MMCs have high demand as structural materials in aerospace, automotive, and railway sectors [7]. However, micron-sized ceramic particulates reduce ductility and generate crack during mechanical loading, further leading to premature structural failure [8,9]. The size of reinforcing particles affects the mode of failure, strength, and ductility of particulate reinforced composites. Mechanical performance of MMCs can be further enhanced by decreasing the size of reinforcing particulates and/or matrix grains from the micrometer to nanometer level [10–12]. Nanoparticles reinforced metal matrix composites can be considered as the second generation of metal matrix composites. Over the past years, there have been increasing interests to produce metal matrix nanocomposites due to their significantly enhanced performances compared to composites with micro-sized reinforcements [13]. For example, the tensile strength of 1 vol. % Si_3N_4 (10 nm)-Al composites is comparable to that of a 15 vol. % SiC (3.5 µm)-Al composite [14]. The addition of 1 wt. % SiC carbide nanoparticles as reinforcements in 356 aluminum alloy could enhance the ultimate tensile strength and yield strength by 100% while the ductility remained almost unchanged [15]. This motivates the development of nanoparticles reinforced metal matrix composites (nano-MMCs) which is one of the rapidly evolving research areas in advanced composites [16–18].

Material removal mechanism during the machining of particle-reinforced MMCs is different from that of monolithic metals. The complex deformation mechanism due to the presence of reinforcements [19,20] in metal matrix composites (MMCs) causes high tool wear during traditional machining. Numerous reports can be found in literature describing experimental, analytical and numerical investigations related to machining of micro-particle reinforced MMCs. Nonetheless, there are very limited investigations on the machining of nanoparticles reinforced MMCs. Li et al. [21] investigated the machinability of magnesium-based MMCs with 5, 10 and 15 vol. % reinforcements of SiC nano-particles (particle diameter: 20 nm) with respect to pure magnesium. The milling was performed using two flutes 1.016 mm cutting tool at different feed rates (i.e., 0.5, 1 and 1.5 mm/s) and spindle speeds (i.e., 20, 40 and 60 krpm). Cutting forces, surface morphology and surface roughness were measured and analyzed using response surface methodology (RSM) to optimize the machining conditions. The cutting force increased with increasing volume fraction of nanoparticles. The influence of a single experimental variable such as, feed rate, spindle speed or nanoparticles volume fraction on surface roughness was not manifested. Liu et al. [22] proposed a machining force model by considering three machining zones such as shearing, ploughing, and elastic recovery during micro-milling of SiC nanoparticles reinforced Mg-MMCs. The volume fraction of particles and particle size were considered as two significant factors affecting the cutting forces in that model. It was found that the amplitude and profile of cutting forces varied with the volume fraction of reinforcing particles owing to the strengthening effect of SiC nanoparticles. Teng et al. [23] investigated the effect of (a) types of reinforcement materials; (b) weight fraction of reinforcements; (c) feed per tooth; (d) spindle speed and (e) depth of cut on cutting force, surface morphology and chip formation during micromilling of Mg-based ZnO and BN nanoparticle reinforced MMCs. It was found that, machining force for pure Mg is larger than that for MMCs except the MMC reinforced with 2.5 wt. % ZnO particles. ZnO particles reinforced MMCs exhibited higher machining force than that of BN particle reinforced MMCs. The chips of 2.5 wt. % BN reinforced MMCs possessed short and tightly curled shape features due to their reduced compressive ductility. Teng et al. [24] studied micromilling of Ti and TiB_2 nanoparticles (volume fraction of 1.98 vol. % and average particle size of 50 nm) reinforced Mg MMCs using AlTiN-coated tool. Abrasive wear and chip adhesion were observed along main cutting edges of the tool. Larger cutting force and worse surface finish were obtained at small feed per tooth ranging from 0.15 to 0.5 µm/tooth, which was indicative of strong size effect. Chip adhesion effect was more evident during the machining of MMCs with nano-sized Ti particles compare to TiB_2 particles, which was associated with the ductile nature of the matrix (Mg) and reinforcement material (Ti). Th analysis of variance suggested that the spindle speed and depth of cut affected the surface roughness significantly.

Thus, based on survey of the literature, it is evident that the investigation on machining of nanoparticle reinforced MMCs is at the early stage and mainly limited to micromachining of magnesium matrix-based MMCs. To further elaborate the current understanding, the current study investigated the conventional milling of SiC nanoparticle reinforced 6061 aluminum alloy in terms of surface finish, chip formation and force generation in order to benefit materials researchers and engineers from the outcomes of this study.

2. Experiments

6061 aluminum matrix MMCs reinforced with SiC particles (Particle size: ~700 nm) were milled in a dry condition. SiC particles were of irregular shape and 10% by volume of MMCs. The 6061 aluminum alloy contains 98.5, 0.7, 0.6 and 0.2 wt. % Al, Mg, Si and Cu respectively [25]. Milling was conducted on a three-axis Leadwell V30 CNC vertical machining center with maximum machine table movements of 760, 410 and 520 mm along the x, y and z axes, respectively. The cutting tool inserts were changed after every experiment to avoid tool wear. The parameters kept constant throughout the experiments are rake angle: 12°, axial depth of cut: 1.5 mm, radial depth of cut: 3 mm, approach angle: 90°, number of tooth: 1, relief angle: 7°, nose radius: 1 mm, tool holder diameter: 12 mm and cutting tool insert: SANDVIK R390-11 T3 08E-ML S40T. The variation of parameters considered in this study is given in Table 1.

Table 1. Parameter variations in milling metal matrix composites (MMCs) reinforced with SiC particles.

Particle Size (nm)	700		
Speed (m/min)	60	100	140
Feed (mm/tooth)	0.025	0.0	0.075

The machining was performed at different levels, as shown in Table 2. When one parameter varied, the other parameter was kept constant. For each experiment, the machining forces and surface roughness were measured. Chips and machined surface were examined under an Olympus SC100 optical microscope (Tokyo, Japan). Chip thickness was measured using a Vernier calliper. The roughness of the machined surface was measured using a portable stylus-type surface profilometer (SJ-201; Mitutoyo Surftest, Washington, DC, USA). Machining forces were obtained by a Kistler dynamometer (Victoria, Australia) while Dynoware28 software was used to provide and evaluate high-performance and real-time graphics for cutting forces.

Table 2. Details of design of experiments.

Experiment No.	Cutting Speed (mm/min)	Feed Rate (mm/tooth)
1	60	0.025
2	60	0.05
3	60	0.075
4	100	0.025
5	100	0.05
6	100	0.075
7	140	0.025
8	140	0.05
9	140	0.075

3. Results and Discussions

3.1. Machined Surfaces

The average surface roughness of nano-MMCs machined at different feeds and speeds is shown in Figure 1 with respect to machining speed and feed rate. The trend of surface roughness changes with the change of machining conditions. When the machining speed is slow (60 mm/min), the roughness also appears to be lower at smaller feed but it increases with the increase of feed. However, at faster machining speeds (100–140 mm/min), surface roughness is higher at lower feed and then it decreases slightly with the increase of feed. Finally, it remains almost constant with a further increase of feed. Figure 1 also demonstrates that the lowest surface roughness is achieved at the smallest feed (0.02 mm/tooth) and slowest speed (60 mm/min).

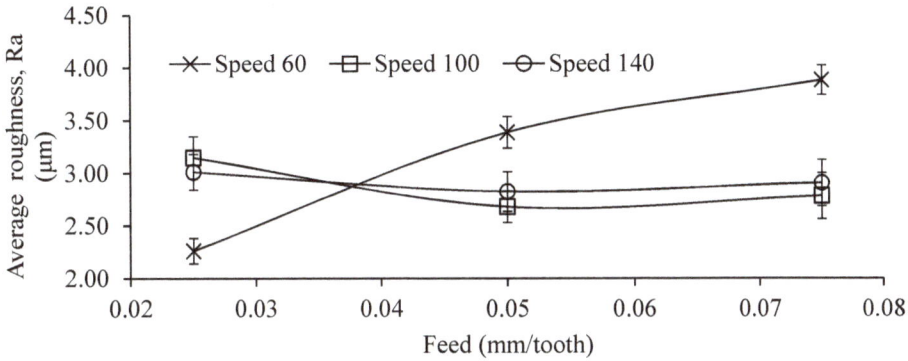

Figure 1. Effect of feed and speed during milling of nanoparticle reinforced MMCs.

The machined surfaces consist of traces of nose edge profiles for cutting tools. The depth and extent of profile depend on the depth of cut and feed. However, material properties such as ductility, brittleness, and elasticity also distort the profile of tool nose edge compared to original shape. These also incorporate additional features such as brittle fracture and side flow on machined surfaces based on material properties. Machining speed and feed rate influence material properties in machining zone. Higher speed increases the strain rate and temperature. On the other hand, higher strain rate works to harden the materials while higher temperature softens it. The structure of materials itself plays an important role in determining hardening or softening effect. For a given length of cut, at the low feed, the distance between two successive tool paths is less and hence a higher number of tool–particle interactions will occur than at the higher feed [20]. However, as the particles are very small, surface damage due to tool-particle interactions is almost absent [23]. Therefore, surface roughness at the lower feed is smallest in case of nano-MMCs. At the higher speed and lower feed, thermal effect dominates, resulting in the increase in the ductility of MMCs. In addition, tensile ductility of nano-MMCs naturally increases with the incorporation of highly elastic nano-sized reinforcements to facilitate elastic recovery [23]. These contribute to higher surface roughness at the higher speed and lower feed. With the feed increase, temperatures decreases due to increased distance between two successive tool paths. This reduces the ductility and allows machined surfaces to replicate the tool path with slightly better surface finish.

The profiles of machined surfaces in different conditions are shown in Figure 2. The peaks and valleys of different magnitudes are clearly visible in all of the surfaces. The most regular and uniform peaks and valleys are noted at the slowest speed and lowest feed. In other cases, the peaks and valleys are irregular. The surface profiles cannot exhibit any effect of reinforcing particle explicitly. The surface profiles support the above-mentioned discussion as ductility and elastic recovery of nano-MMCs vary at different machining feeds and speeds.

The morphology of machined nano-MMC surfaces in different machining conditions is presented in Figure 3. The traces of cutting tools are clearly visible in all cases in addition to the material flow. The material flowed in a way that lobes and layers were generated. All the machined surfaces got pitting dots and various sizes of holes/craters. Extended lobes, smaller craters and fewer layers were noted on surfaces machined at the low speed and low feed (Figure 3a). Such phenomena suggested that the material had better ductile flow in this condition. With the increase of feed, traces of cutting tools were intersected as shown in Figure 3b. In this case, shorter lobes at different directions were evident. This demonstrated less ductile material behavior at the higher feed and lower speed. When the speed is high and feed is low, the machined surface contained bigger crates almost with no lobe, as shown in Figure 3c. This indicates brittle material behavior in such a machining condition. At the higher speed and feed, machined surface also contained lots of crates and minor lobes for material deformation (Figure 3d). In this case, the material is neither so ductile nor so brittle. Therefore the average surface roughness varies accordingly with the change of corresponding machining conditions.

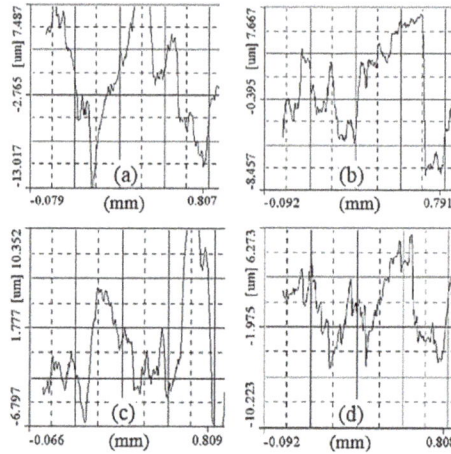

Figure 2. Surface profile of machined surface at: (**a**) 60 mm/min speed, 0.025 mm/tooth feed; (**b**) 60 mm/min speed, 0.075 mm/tooth feed; (**c**) 140 mm/min speed, 0.025 mm/tooth feed and (**d**) 140 mm/min speed, 0.075 mm/tooth feed.

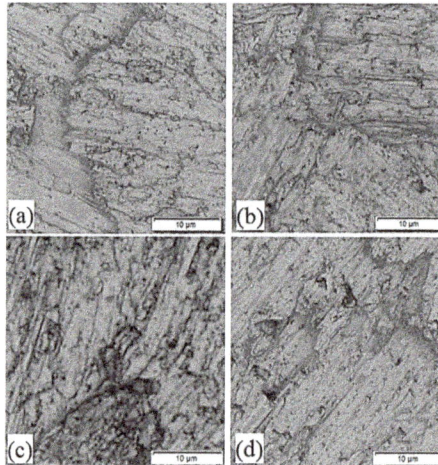

Figure 3. Morphology of the machined surfaces in different conditions: (**a**) 60 mm/min speed, feed 0.025 mm/tooth; (**b**) 60 mm/min speed, 0.075 mm/tooth feed; (**c**) 140 mm/min speed, 0.025 mm/tooth feed and (**d**) 140 mm/min speed, 0.075 mm/tooth feed.

The overlaps of milling tool path were observed on machined surfaces due to repeated cutting on machined surfaces by the cutting tool. During machining, a portion of material was pushed by the cutting edge and then elastically recovered to its original position after the tool pass instead of being removed plastically. Consequently, machined surfaces were cut repeatedly. The tensile ductility of nano-MMCs was improved because of highly elastic nano-sized reinforcements to facilitate elastic recovery [23]. During the machining of traditional MMCs with micro-size particles, void marks could be left on machined surfaces due to particle pull-out [26] and thus deteriorate the surface finish.

However, particle pull-out and fracture were not noticed on machined nano-MMC surfaces. The lower surface roughness for nano-MMCs compared to that for MMCs with micro-size particles revealed a relatively good machinability of such materials [23].

3.2. Machined Chips Morphology

The ratio of uncut chip thickness (feed/rev) to thickness of chips is known as chip thickness ratio or simply the chip ratio. This term gives information on chip formation and material deformation behavior in the machining zone during the material removal process. During metallic material milling, chips are formed along shear plane and they slide on the rake face of the cutting tool. The chip flow of metal is shorter and thicker than the metal prior to cutting because of plastic deformation at share plane/zone. Friction and/or adhesion with the rake face slow down the flow rate of upward chips compare to cutting speed. The smaller chip ratio indicates that the chips undergo more plastic deformation and vice versa. Figure 4 show the chip ratio of nano-MMCs after machining at different feeds and speeds. Chip ratio increases with the increase of feed or uncut chip thickness for all cutting speeds. Chip ratio at the lowest speed of 60 m/min is smaller than that at the higher speeds in the case of all feeds. Chip ratio increases with the increase of speed at 100 m/mm and then decreases again with the further increase of speed.

At a higher feed, the uncut chip thickness is larger. Sever plastic deformation in the chips takes place at tool-chip interfaces. However, the degree of plastic deformation decreases as the distance from tool-chip interface increases towards the top of chips. This causes relatively thin chips resulting in a higher chip ratio. As mentioned previously, machining speed contributes to material softening due to heat generation and material hardening by the induced strain rate. These two opposite effects are traded in at different levels based on material structures. When softening effect takes over, it is easy for the chips to slide over the tool rake face and elongate in the longitudinal direction instead of thickening. Therefore, the chips become relatively thinner and chip ratio is much higher. An opposite effect takes place when relative chip thickness increases due to strain hardening effect. It seems that, for nano-MMCs considered in this investigation, at lower speed, the strain hardening takes over the softening effect. However, with the increase of speed, the contribution of softening effect is more pronounced. With the further increase of speed, the contributions of the hardening effect become dominant and therefore the chip ratio varies accordingly.

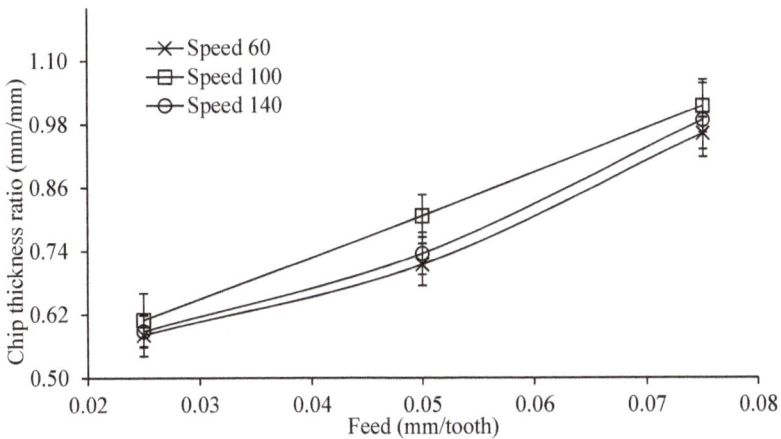

Figure 4. Chip thickness ratio of nanoparticles reinforced with Al-based metal matrix composites (nano-MMCs) at different feeds and speeds.

The top surfaces of chips formed under different machining conditions are shown in Figure 5. It is evidently shown that the chips are full of wrinkles. These wrinkles are generated due to curly chips. Uneven strain across the plastic zone during the chip formation causes curly chips, which in turn depends on the ductility/brittleness of workpiece materials. Brittle materials generate chips will little or no curl contrary to ductile materials which shows the formation of long spiral chips with wrinkles [26]. Therefore, nano-MMCs retain most of the ductility of matrix materials due to nano-sized reinforcements.

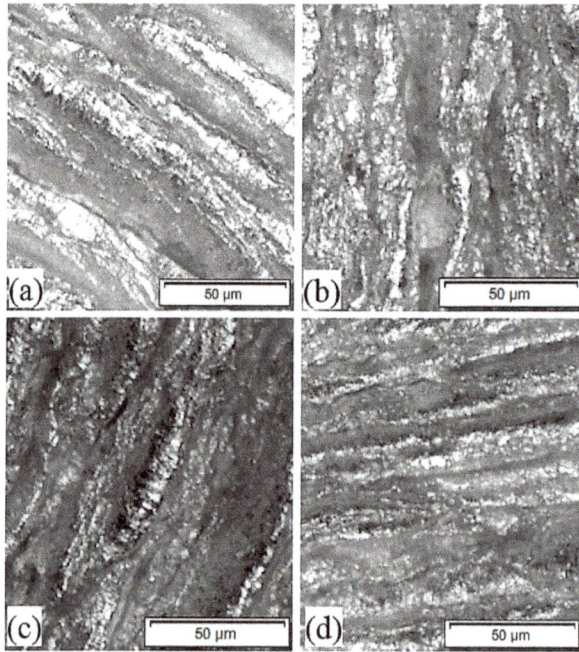

Figure 5. Top surfaces of chips in different conditions: (**a**) 60 mm/min speed, 0.025 mm/tooth feed; (**b**) 60 mm/min speed, 0.075 mm/tooth feed; (**c**) 140 mm/min speed, 0.025 mm/tooth feed and (**d**) 140 mm/min speed, 0.075 mm/tooth feed.

Figure 6 shows the bottom surfaces of the chips at different machining conditions. Apparently, bottom surfaces of chips got quite thicker layers of deformed materials (Figure 6b,d) due to their material flow at the higher feed irrespective of speed. The surfaces machined at lower feed show thin layers of deformed materials. Though the chip surfaces are very similar at both higher and lower speeds (Figure 6c,d), a close look reveals that the surface generated at higher milling speed is smoother owing to thermal softening of materials at the higher speed. It seems that at a lower feed the extended friction is more dominant and at higher feed both adhesion and material flow occur at tool-chip interface. Therefore, the chip ratio is altered accordingly.

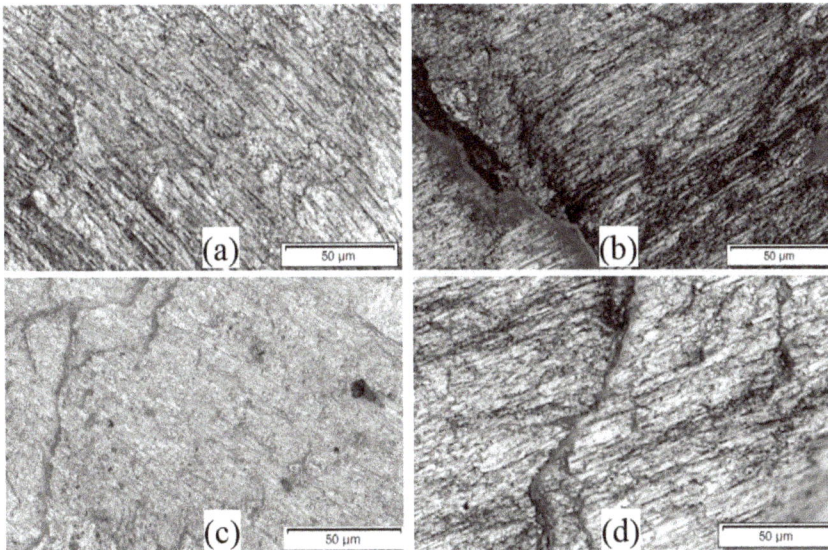

Figure 6. Bottom surfaces of chips in different conditions (**a**) 60 mm/min speed, 0.025 mm/tooth feed; (**b**) 60 mm/min speed, 0.075 mm/tooth feed; (**c**) 140 mm/min speed, 0.025 mm/tooth feed and (**d**) 140 mm/min speed, 0.075 mm/tooth feed.

3.3. Machining Forces

The machining forces at radial (F_x), cutting/tangential (F_y) and thrust directions during the milling of nano-MMCs are shown in Figure 7. Highest force was measured in the cutting direction, smallest force was measured in radial direction and thrust force was in between them. These trends did not change with the variation of machining conditions. At all speeds, all the forces increased with the increase of feed since more materials were removed at the higher feed. The variation of forces with the speed change was not significant at 0.025 and 0.05 mm/tooth feed and minor variations did not follow any trend. However, the forces decreased with the increase of speed at 0.075 mm/tooth feed. It appears that at lower feeds, contribution of softening effect of the speed is not significant to reduce machining forces. However, the softening of materials due to the thermal effect takes place with the increase of speed at the high feed rate. The signals of forces at different machining conditions are presented in Figure 8. It showed that in all cases the forces started to increase as the tool cutting edge engaged to remove materials. After that initial engage, the forces were maximized when the maximum material thickness was cut and finally decreased as the cut thickness was reduced to the minimum level. At the end of cutting cycle, the cutting tool was pulled towards the workpiece just before the disengagement. This gives a portion of negative forces in the cutting cycles in order to achieve the maximum quantity for radial force (F_x).

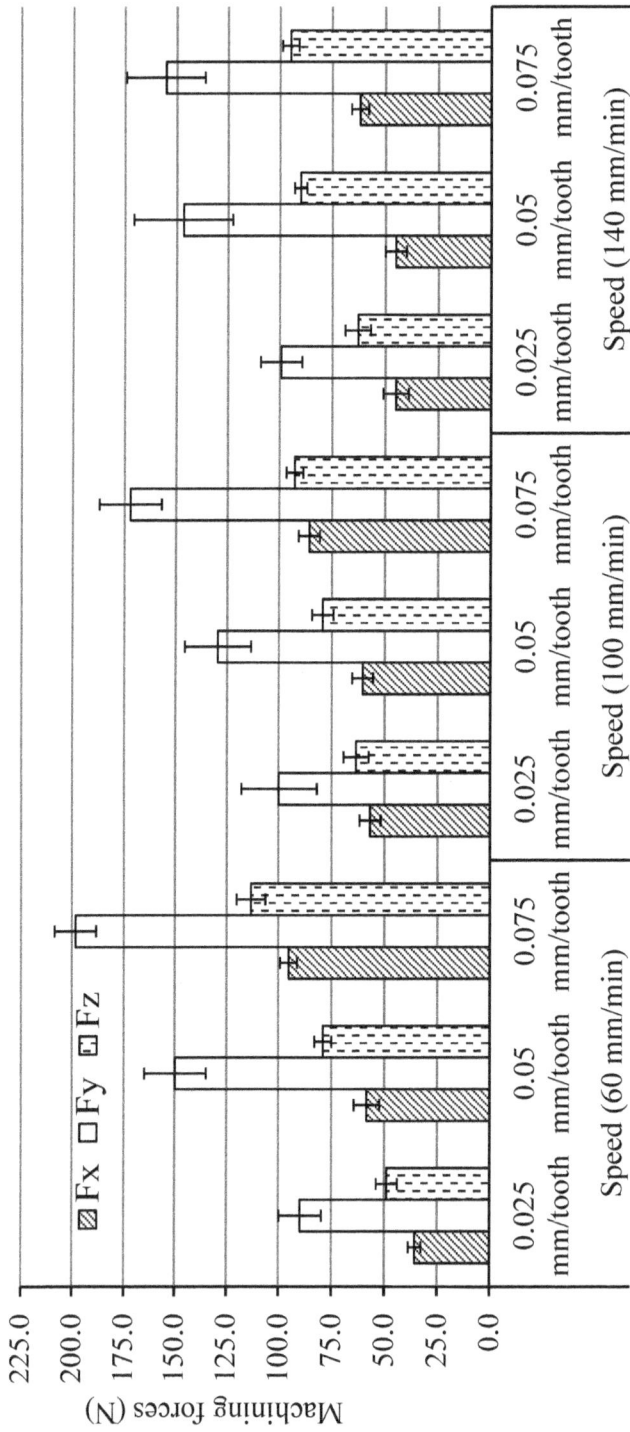

Figure 7. Machining forces in different conditions.

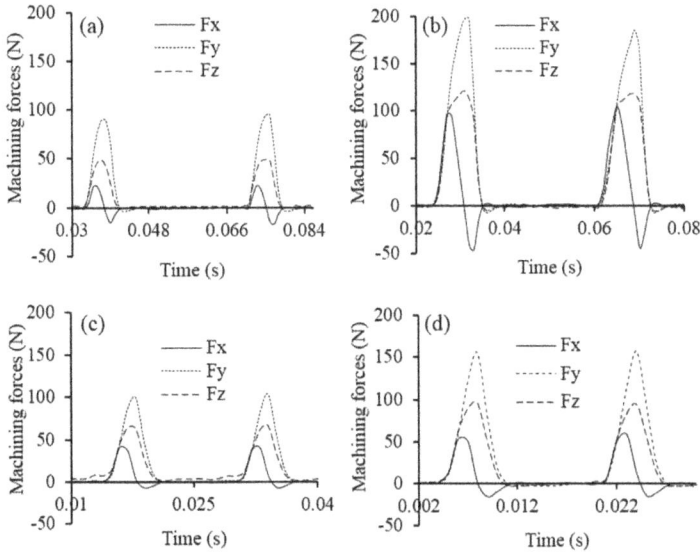

Figure 8. Force signals in different conditions: (**a**) 60 mm/min speed, 0.025 mm/tooth feed; (**b**) 60 mm/min speed, 0.075 mm/tooth feed; (**c**) 140 mm/min speed, 0.025 mm/tooth feed and (**d**) 140 mm/min speed, 0.075 mm/tooth feed.

4. Conclusions

The present investigation explored the behavior of SiC nanoparticles reinforced with Al-based MMCs under different milling conditions. The following conclusions can be drawn based on the results presented:

1. Under different milling conditions, the surface generated on nano-MMCs is similar to that of monolithic materials due to the absence of particle fracture and pull-out related damage. Surface defects, such as lobes and layers, were minimum and the average surface roughness increased with the increase of feed at the lower speed. The effect of feed was negligible at higher speeds.
2. The chip ratio was higher at the higher feed for all cutting speeds. Wrinkles were noted on the top surface of the chips due to curly shape. At lower feed, the friction is more dominant in contrast to higher feed scenario where both adhesion and materials flow contribute to the chip formation.
3. Machining forces increase with the increase of feed at all speeds with the increase of material removal rate. The influence of machining speed on the forces was insignificant without following any trend. These behaviors could be attributed to the combined effects of thermal softening and strain hardening of machined materials in different machining conditions.

Author Contributions: Alokesh Pramanik and Animesh Basak designed and performed the experiments; Yu Dong, Subramaniam Shankar and Guy Littlefair contributed reagents/materials/analysis tools and write the paper.

Conflicts of Interest: The authors declare no conflict of interest.

References

1. Basak, A.; Pramanik, A.; Islam, M.N. Failure mechanisms of nanoparticle reinforced metal matrix composite. *Adv. Mater. Res.* **2013**, *774–776*, 548–551. [CrossRef]
2. Basak, A.; Pramanik, A.; Islam, M.N.; Anandakrishnan, V. Challenges and recent developments on nanoparticle-reinforced metal matrix composites. *Fill. Reinf. Adv. Nanocompos.* **2015**, *14*, 349–367. [CrossRef]
3. Basumallick, A.; Sarkar, J.; Das, G.; Mukherjee, S. On the Synthesis of In Situ Ni–SiO$_2$ Nanocomposites by Isothermal and Non Isothermal Reduction Technique. *Mater. Manuf. Process.* **2006**, *21*, 648–651. [CrossRef]
4. Ferkel, H.; Mordike, B. Magnesium strengthened by SiC nanoparticles. *Mater. Sci. Eng. A* **2001**, *298*, 193–199. [CrossRef]
5. Hassan, S.; Gupta, M. Development of high performance magnesium nano-composites using nano-Al$_2$O$_3$ as reinforcement. *Mater. Sci. Eng. A* **2005**, *392*, 163–168. [CrossRef]
6. Huda, D.; El Baradie, M.A.; Hashmi, M.S.J. Analytical study for the stress analysis of Metal Matrix Composites. *J. Mater. Process. Technol.* **1994**, *45*, 429–434. [CrossRef]
7. Kang, Y.-C.; Chan, S.L.-I. Tensile properties of nanometric Al$_2$O$_3$ particulate-reinforced aluminum matrix composites. *Mater. Chem. Phys.* **2004**, *85*, 438–443. [CrossRef]
8. Li, J.; Liu, J.; Liu, J.; Ji, Y.; Xu, C. Experimental investigation on the machinability of SiC nano-particles reinforced magnesium nanocomposites during micro-milling processes. *Int. J. Manuf. Res.* **2013**, *8*, 64–84. [CrossRef]
9. Liu, J.; Li, J.; Ji, Y.; Xu, C. Investigation on the effect of SiC nanoparticles on cutting forces for micro-milling magnesium matrix composites. In Proceedings of the ASME 2011 International Manufacturing Science and Engineering Conference, Corvallis, OR, USA, 13–17 June 2011.
10. Lü, L.; Lai, M.; Liang, W. Magnesium nanocomposite via mechanochemical milling. *Compos. Sci. Technol.* **2004**, *64*, 2009–2014. [CrossRef]
11. Ma, Z.; Li, Y.; Liang, Y.; Zheng, F.; Bi, J.; Tjong, S. Nanometric Si$_3$N$_4$ particulate-reinforced aluminum composite. *Mater. Sci. Eng. A* **1996**, *219*, 229–231. [CrossRef]
12. Miracle, D. Metal matrix composites–from science to technological significance. *Compos. Sci. Technol.* **2005**, *65*, 2526–2540. [CrossRef]
13. Paramsothy, M.; Chan, J.; Kwok, R.; Gupta, M. Al$_2$O$_3$ nanoparticle addition to commercial magnesium alloys: Multiple beneficial effects. *Nanomaterials* **2012**, *2*, 147–162. [CrossRef] [PubMed]
14. Pramanik, A.; Arsecularatne, J.; Zhang, L. Machining of particulate-reinforced metal matrix composites. In *Machining*; Springer: New York, NY, USA, 2008; pp. 127–166.
15. Pramanik, A.; Basak, A.K. Fracture and fatigue life of Al-based MMCs machined at different conditions. *Eng. Fract. Mech.* **2018**, *191*, 33–45. [CrossRef]
16. Pramanik, A.; Basak, A.K. Effect of machining parameters on deformation behaviour of Al-based metal matrix composites under tension. *Proc. Inst. Mech. Eng. B J. Eng. Manuf.* **2018**, *232*, 217–225. [CrossRef]
17. Pramanik, A.; Littlefair, G. Fabrication of nano-particle reinforced metal matrix composites. *Adv. Mater. Res.* **2013**, *651*, 289–294. [CrossRef]
18. Pramanik, A.; Zhang, L.; Arsecularatne, J. Micro-indentation of metal matrix composites—An FEM analysis. *Key Eng. Mater.* **2007**, *340–341*, 563–570. [CrossRef]
19. Pramanik, A.; Zhang, L.; Arsecularatne, J. Machining of metal matrix composites: Effect of ceramic particles on residual stress, surface roughness and chip formation. *Int. J. Mach. Tools Manuf.* **2008**, *48*, 1613–1625. [CrossRef]
20. Pramanik, A.; Zhang, L.C.; Arsecularatne, J.A. Deformation mechanisms of MMCs under indentation. *Compos. Sci. Technol.* **2008**, *68*, 1304–1312. [CrossRef]
21. Teng, X.; Huo, D.; Wong, E.; Meenashisundaram, G.; Gupta, M. Micro-machinability of nanoparticle-reinforced Mg-based MMCs: An experimental investigation. *Int. J. Adv. Manuf. Technol.* **2016**, *87*, 2165–2178. [CrossRef]
22. Teng, X.; Huo, D.; Wong, W.L.E.; Sankaranarayanan, S.; Gupta, M. Machinability Investigation in Micro-milling of Mg Based MMCs with Nano-Sized Particles. In *Magnesium Technology*; Springer: New York, NY, USA, 2017; pp. 61–69.
23. Tjong, S.C.; Ma, Z. Microstructural and mechanical characteristics of in situ metal matrix composites. *Mater. Sci. Eng. R Rep.* **2000**, *29*, 49–113. [CrossRef]

24. Wang, Z.; Chen, T.K.; Lloyd, D.J. Stress distribution in particulate-reinforced metal-matrix composites subjected to external load. *Metall. Trans. A* **1993**, *24*, 197–207. [CrossRef]
25. Xiaochun, L.; Yang, Y.; Weiss, D. Theoretical and experimental study on ultrasonic dispersion of nanoparticles for strengthening cast Aluminum Alloy A356. *Metall. Sci. Technol.* **2013**, *26*, 12–20.
26. Zhang, Z.; Zhang, L.; Mai, Y.W. Modeling steady wear of steel/Al_2O_3-Al particle reinforced composite system. *Wear* **1997**, *211*, 147–150. [CrossRef]

© 2018 by the authors. Licensee MDPI, Basel, Switzerland. This article is an open access article distributed under the terms and conditions of the Creative Commons Attribution (CC BY) license (http://creativecommons.org/licenses/by/4.0/).

Journal of
composites science

MDPI

Article

Manufacturing and Mechanical Properties of Graphene Coated Glass Fabric and Epoxy Composites

Rehan Umer [1,2]

[1] Centre for Future Materials, University of Southern Queensland, Toowoomba, QLD 4350, Australia;
    rehan.umer@usq.edu.au
[2] Department of Aerospace Engineering, Khalifa University of Science and Technology, Abu Dhabi 127788,
    United Arab Emirates

Received: 6 March 2018; Accepted: 18 March 2018; Published: 21 March 2018

Abstract: The processing characteristics and mechanical properties of glass fabric reinforcements coated with graphene nanoparticles were investigated. Graphene was coated onto either one or both sides of a plain weave glass fabric. The coated fabrics were investigated to measure key process characterization parameters used for vacuum assisted resin transfer molding (VARTM) process which are, reinforcement compaction response, in-plane, and transverse permeability. It was found that graphene coated glass reinforcements were stiffer than the pure glass reinforcements which will have direct influence on final fiber volume fraction obtained during VARTM processing. The permeability measurement results show that the graphene coated reinforcements filled relatively slower compared with the pure glass samples. Composite samples were then tested for flexural and low velocity impact. The initial results show that the flexural modulus did not change as the wt % of graphene increases. However, a decrease in flexural strength with increasing wt % of graphene was observed. It was also observed that the coating of graphene on glass reinforcements caused delamination between plies and resisted localized damage under low velocity impact as compared to pure glass samples.

Keywords: graphene coatings; VARTM; mechanical properties; impact response

1. Introduction

Graphene is the basic structural unit of some carbon allotropes including graphite, carbon nanotubes, and fullerenes with promising mechanical, electrical, optical, thermal and magnetic properties [1–4]. Recent progress has shown that graphene-based materials can have a profound impact on electronic and optoelectronic devices, chemical sensors, nanocomposites and energy storage [5–9]. The addition of exfoliated graphite nanoplatelets to the polymer matrix has been shown to produce nanocomposites that are multifunctional and significantly improve many of the mechanical properties [10–15]. However, the use of exfoliated graphite nanoplatelets as a secondary reinforcement in glass fiber composite laminates has not been studied extensively. The stiffness, toughness, and wear performance of the composites are extensively determined by the size, shape, volume content, and especially the dispersion homogeneity of the particles. Nanostructure materials provide opportunities to explore new fracture behavior and functionality beyond those found in conventional materials. Ávila et al. [16] showed that failure mechanisms of laminated composites can be influenced by nanostructures formed by nanoparticles dispersed into epoxy systems. According to them, the presence of nanoclay into fiber glass/epoxy composites lead to a more intense formation of delaminated areas after a low-velocity impact test. This phenomenon was attributed to interlaminar shear forces caused by the intercalated nanostructures inside the epoxy system. Furthermore, the energy absorption of these laminates increased by 48% with dispersion of 5 wt % of nanoclays.

Vacuum Assisted Resin Transfer Molding (VARTM) has proven to be a robust, low-cost technique for the manufacture of composite structures. Recent studies on the effects of nanoparticles on resin infiltration during manufacture of composites by VARTM have focused primarily on carbon nanotubes, carbon nanofibers [17] and nanoclay [18]. Little information has been published on VARTM processing effects using graphite nanoplatlets. There are two primary methods used to introduce nanoparticles into composite materials. One method involves dispersing the nanoparticles in the resin [19–23]. However, the high surface area and aspect ratio of the nanoparticles can result in an increase in resin viscosity [17]. Furthermore, during resin infiltration, aggregation of the nanoparticles can occur within the fiber tows. A second technique is to coat the nanoparticles directly onto the fibers [24–26] which eliminates the problems observed with the first technique.

In this study the effects of exfoliated graphite nanoplatelets on the processing characteristics and mechanical properties of glass fabric composites fabricated by the VARTM process were investigated. It is anticipated that the addition of the nanoplatelets will improve the out-of-plane properties of the composite including the interlaminar strength and fracture toughness and damage due to impact [19,27,28]. The large surface area of exfoliated graphite nanoplatelets is one of the most attractive characteristics of this kind of nanoparticles, which facilitates creating a large interface area in a nanocomposite.

In this investigation, exfoliated graphite dispersions will be coated onto the surface of glass fabric reinforcements. The compaction characteristics and permeabilities of the nanoparticle coated glass fabrics will be measured to determine the impact of nanoparticles on the processability under VARTM conditions. Composite structures containing nanoplatelets will be subjected to mechanical tests to study the influence of nanoplatelets on mechanical performance and fracture behavior under impact loading.

2. Experimental Section

2.1. Materials

An 800 g/m^2 plain weave glass fabric supplied by Owens Corning was used as reinforcement material. A two part toughened epoxy system, Applied Polymeric SC-15 epoxy resin and SC-15 amine hardener with an ambient mixed viscosity of 0.30 Pa.s was used as a matrix. Hydraulic oil with similar viscosity as the resin system was used for permeability experiments. The nano-reinforcement was exfoliated graphite nanoplatelets. The diameter of the nanoplatelet was approximately 5–6 μm and the thickness was approximately 7–8 nm.

2.2. Graphene Coating Process

As a first step, the exfoliated graphite nanoplatelets were further processed to break the agglomerations to form graphene. A solution of approximately 5% graphite by weight mixed in 2-isopropanol was prepared. The solution was mixed using both a mechanical stirring device and a sonicator, Figure 1a. The 2-isopropanol and graphite mixture was stirred for 30 min followed by sonication at 35 W power for 2 h with on/off pulses. The glass fabric was cut to the desired dimensions and weighed accordingly. The measured amounts of graphene in 2-isopropanol solution was then brushed onto the glass fabric until it was evenly distributed as shown in Figure 1b. The coated fabrics were placed beneath a fume hood for 24 h until the 2-isoproponal was evaporated. Samples were prepared with 0.5 wt % and 1.0 wt % graphene coated on one surface of the glass fabric and with 0.5 wt % coated on both sides of the glass fabric. Samples were also prepared by coating the surface of the glass fabric with 2-proponal to assess the effect of the solvent on processing and overall properties. These samples are referred to as 0 wt % graphene. The results were also compared to "As Received" pure glass samples.

Figure 1. (**a**) Preparation of graphene solution through ultrasonication; (**b**) graphene coating process onto plain weave glass fabric.

2.3. Compaction Characterization

To assess the process-ability of the coated fabrics under VARTM pressure, the coated reinforcements were subjected to compaction experiments conducted under dry conditions to measure the compaction response of the glass preforms. The compaction test fixture is composed of two flat steel plates used to compact the samples. The upper and lower platen has dimensions of 15 cm × 15 cm. A laser displacement sensor (L-Gage, Banner, LG10, Minneapolis, MN, USA) and a digital dial indicator were used to monitor the crosshead displacement and thus the thickness of the sample being tested. A photo of the compaction test setup is shown in Figure 2. The fixture was mounted between upper and lower platens of an MTS Insight Material Testing Machine attached with 100 kN load cell. The output data of the L-Gage sensor was gathered using LabVIEW data acquisition software (National Instruments, Austin, TX, USA). Before using the L-Gage sensor, it must be calibrated by defining the minimum and maximum distance limits or distance range.

Figure 2. Picture of the compaction test setup.

In compaction characterization experiments, a compressive load was applied up to a set high load limit and then releasing this load down to a set low load limit. The test cycle was decomposed into two cycles: Loading and Unloading. Both cycles were performed with a constant crosshead speed of 0.5 mm/min. The user inputs to the test program were: high load limit, low load limit, crosshead speed, and data acquisition frequency. The MTS machine load cell was calibrated first. After the calibration, the cross head with upper plate attached was lowered all the way down until it touched the bottom plate. At this point, the extension reading of the machine was zeroed. The dial indicator was also placed on the crosshead and zeroed at this position. Then, the crosshead was raised and the sample to be tested was placed on the bottom plate. The crosshead was manually and slowly lowered until the upper plate touched the sample and a minimal initial load (approximately 10 N) was attained. This indicates the initial thickness of the preform. Glass preform specimens were compacted to 101.5 kPa which corresponds to the maximum VARTM compaction pressure. Once the maximum load limit was reached, the crosshead begins unloading the specimen until the load drops to zero.

2.4. Permeability Characterization

Separate fixtures were used to measure the in-plane (K_{11} and K_{22}) and the transverse (K_{33}) permeabilities. The transverse or through-thickness permeability test fixture was designed to establish one-dimensional saturated flow of fluid through the preform and is shown in Figure 3. This fixture was designed to accommodate 100 mm diameter preform specimens. The fluid was injected through the thickness of the specimen by rigid distribution plates mounted in the plunger and in the bottom of the cavity. The plates were machined with 2 mm holes drilled in round patterns. A single linear voltage differential transducer (LVDT) was used to measure the thickness of the preform specimen. Two pressure transducers were located at the inlet and outlet to measure the pressure gradient in the transverse direction.

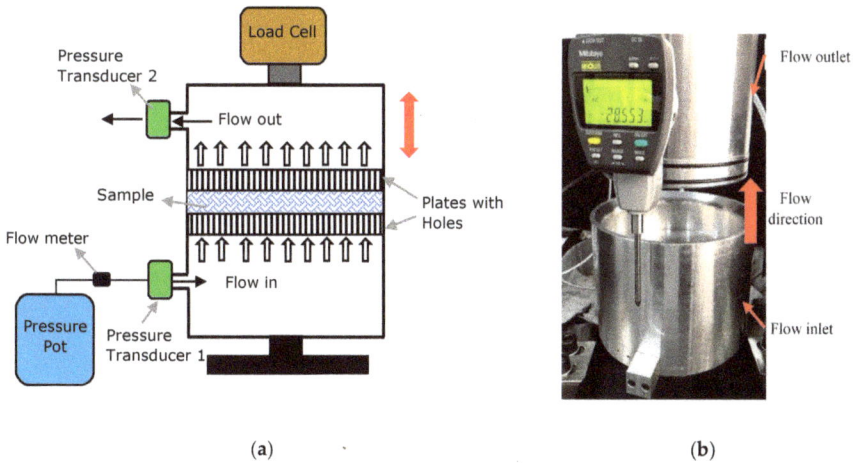

(a)   (b)

Figure 3. (a) Schematic diagram of Transverse Permeability Fixture; and (b) photograph of the fixture.

The preform specimens were placed inside the cavities of permeability fixtures. Once the crosshead was lowered to the desired starting thickness or fiber volume fraction level, the test fluid was injected into the mold cavity under 1 bar pressure. A mass balance was used to measure the flow rate of the test fluid at the outlet. Once steady-state flow conditions were established, the inlet and outlet pressures were measured. At each fiber volume fraction, the difference between inlet and outlet pressures over a range of different flow rates was measured and the data was used to construct a curve of volumetric flow rate versus the pressure drop. Measuring the slope of the curve gives the average permeability for the preform at the specified fiber volume fraction.

The in-plane permeabilities of the graphene coated glass fabrics could not be measured using a one-dimensional flow of fluid through a saturated preform due to possible washout of the nanoparticles observed during the measurements. Hence, a transient or advancing front measurement technique was used, where the flow front positions are recorded as a function of time. The permeability fixture shown in Figure 4 was used to obtain the measurements and includes a clear glass plate as top platen and a rigid steel bottom plate with inlet and outlet holes. Mounted above the glass mold was an HD camera which was used to observe the resin flow along the top surface of the preform, as can be seen in Figure 4.

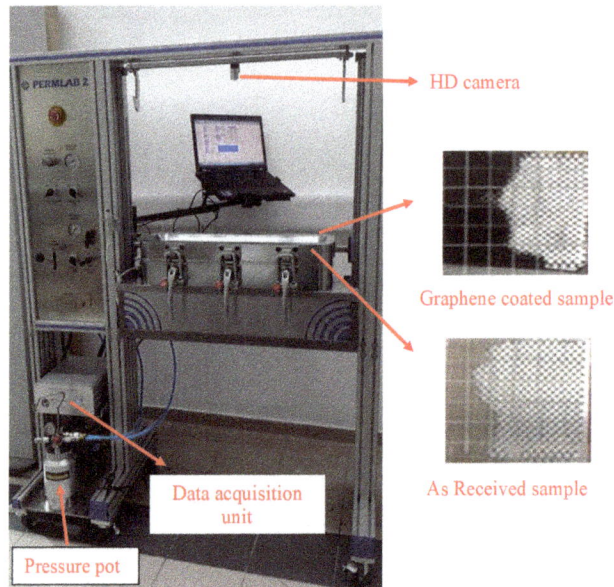

Figure 4. In-plane permeability experimental setup.

Permeability tests were done at each principal in-plane direction i.e., K_{11} and K_{22} by cutting the glass fabric in two principal directions. In-plane permeability tests were performed for samples having different wt % of graphene coatings. Four layers of 400 mm × 100 mm representing 3 mm cavity thickness of glass fabric samples were placed on the tool. A constant injection pressure of 1 bar was used to inject the test fluid inside the mold cavity and flow front positions were monitored throughout the mold filling period.

2.5. Composite Manufacture

The coated glass fabric preforms with dimensions 400 mm × 100 mm with 4 layers thick were infused using two part epoxy resin by the VARTM process. The specimens were fabricated on a glass mold with a line injection port and a line vacuum port. Two layers of resin distribution medium was used. After the materials were placed, the mold was sealed using a vacuum bag and sealant tape. The mold was then infused under vacuum with a pressure of 1 atm. The panel was cured at room temperature for 24 h and post cured in an oven at 70 °C for 7 h. After the cure, test specimens were cut for mechanical testing.

2.6. Mechanical Properties

2.6.1. Flexural Testing

Four point flexural test specimens were prepared in accordance with ASTM D-6272. A third span loading configuration was used as shown in Figure 5. The tests were carried out on an MTS Insight 100 Material Testing Machine (Eden Prairie, MN, USA), with a laser extensometer to accurately measure the deflection. The test fixture had adjustable supports and loading bars that were 6.35 mm in diameter. The test specimens were carefully placed on the support bars to ensure that the loading was symmetric and that the sample was level. The loading rate was based on support span and specimen thickness in accordance with the ASTM standard. The coupons were 13 mm by 60.0 mm with an average thickness of 2.7 mm. The load span was taken as 1/3 of the support span and was according to 16:1 span-to-thickness ratio. The data was recorded at a sampling rate of 2.5 Hz. An average pre-load force of 20 N was applied to start consistently from the same load. The specimens were tested until failure. After testing, the data was analyzed and flexural modulus and strength were calculated. Following equations were used for calculating stress, strain and Young's modulus,

$$\sigma_{max} = \frac{PL}{bd^2} \tag{1}$$

$$E = \frac{0.21 \, mL^3}{bd^3} \tag{2}$$

$$\varepsilon = \frac{4.70 \, Dd}{L^2} \tag{3}$$

where, σ_{max} is the maximum stress, ε is the strain, E is the flexural modulus, P is the load, b is the beam width, d is the thickness, D is the maximum deflection of the center of the beam, L is the support span, and m is the slope of the tangent.

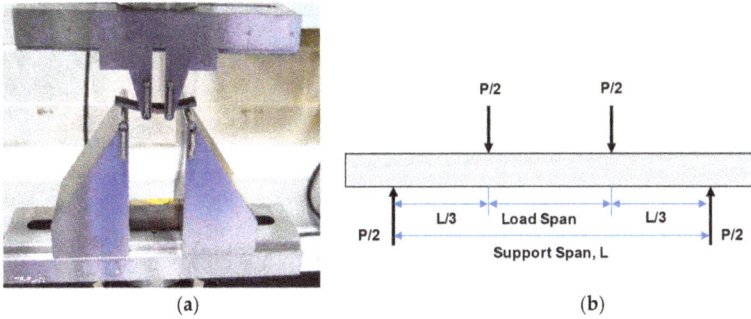

Figure 5. Four point flexural test (**a**) test fixture with sample; (**b**) schematic of the one third of the support span test setup according to ASTM D6272.

2.6.2. Low Velocity Impact

Instrumented Drop-Weight Instron® Dynatup 9250HV impact machine setup was used to test the samples under low velocity impact (2.3–4.8 m/s) loading. The machine setup consists of an instrumented impactor (12 mm diameter) mounted on a crosshead with a provision for attachment of varying weights. The crosshead slides along stiff, smooth guide columns. The specimen is clamped at the base of the machine in a fixture that has circular support. Sample sizes of 10 cm × 10 cm were used for the test. Energy of impact was varied by varying the drop height. The mass was kept constant at 7 kg. The samples were subjected to impact at four different energy levels 20, 40, 60, and 80 J. Load vs. time curves were obtained and displacement and energies were computed.

2.6.3. Ultrasonic Non-Destructive Evaluation

The ultrasonic inspection of the laminate was carried out using a ultrasonic pulser receiver unit (Ultrapac II system with UTwinTM software by Mistras, Princeton Junction, NJ, USA). The scanning was done in pulse–echo immersion mode using a 10 MHz 6.35 mm point focus sensor. In ultrasonic inspection, using the pulse–echo immersion mode, the sample is placed in a water tank and the transducer is brought over the sample. As the ultrasound propagates through the water medium, part of it gets reflected back from the top surface of the sample which is called as front surface echo, while the rest of it passes through the material. The part of ultrasound that is propagating through the sample gets reflected back at the other end of the sample which is called as back surface echo. If there is any defect in the path of the travel of ultrasound, then it acts as reflector and a defect echo is obtained. Therefore, by collecting the information from the back surface echo of ultrasound from the entire surface area of the sample, we can obtain the mapping of the defect in the sample, which is referred to as C-scan. This is done by setting an electronic gate on the back surface echo and digitizing the signal. Such scanning will give the information of cumulative damage as projected onto a horizontal plane. It is possible to set multiple gates from the front surface echo to the back surface echo and collect the information at different interfaces. For the samples subjected to impact loading, scanning was carried out with the impacted surface facing the sensor. The digitized data is further analyzed by pseudo-coloring to get a colored map to differentiate a defective area from the good area.

3. Results and Discussion

3.1. Compaction

Figure 6 presents thickness change as a function of compaction pressure of the glass fiber reinforcements coated with different wt % of graphene. The figure presents both loading and unloading curves for all samples tested. The results were also compared with "As Received" glass reinforcement. The figure shows a clear trend of thickness variation at a set maximum pressure (corresponding to full vacuum condition). As graphene amount increases, the resistance to compaction of the glass reinforcement increases. It is suspected that the solvent affected the properties of the sizing present on glass fibers, causing it to become stiffer. In addition, the nanoparticles fill the voids between the glass tows and hence causing resistance to slipping of fibers under compaction loads. This implies that a glass fiber reinforcement coated with higher wt % of graphene will have higher resistance to compaction and hence the final product will be thick with low glass fiber volume content compared to a product manufactured using "As Received" glass reinforcements.

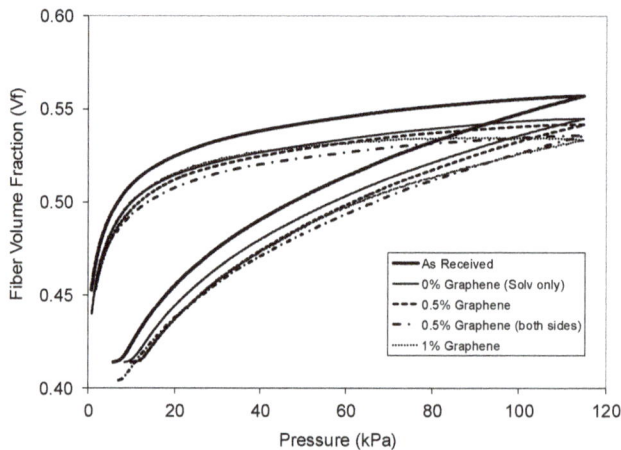

Figure 6. Compaction response of different percentages of graphene coated glass fiber reinforcement.

3.2. Permeability

Permeability characterization tests were performed in three principal directions (K_{11}, K_{22}, and K_{33}) directions. The transverse permeability (K_{33}) test results for glass reinforcements coated with different wt % of graphene are presented in Figure 7. Three sets of tests were completed and the exponential equation was fit to the average data points. The transverse permeability values for glass fabric with graphene 0.5 wt % were very similar. The difference was found in the solvent treated samples and 1 wt % graphene coated samples, where the permeability was found to be higher and lower respectively. It is suspected that due to the use of solvent, the sizing on the glass fibers may have dissolved, as a result leaving behind channels for the fluid to flow through the thickness direction without much resistance. As oppose to samples coated with 1 wt % graphene, where excessive coating left open channels blocked for fluid to flow, resulting in lower permeability.

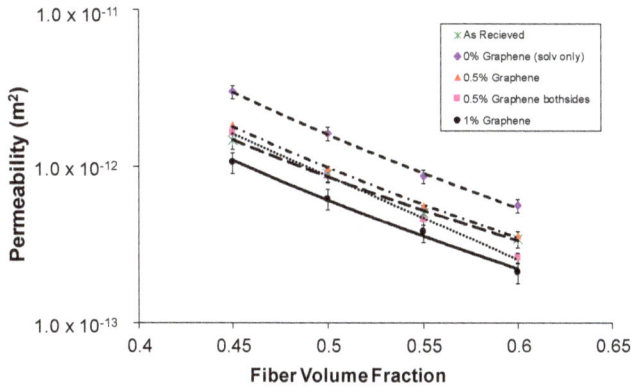

Figure 7. Transverse permeability results of different percentages of graphene coated glass fiber reinforcement.

To measure the in-plane transient permeability (K_{11} and K_{22}), the flow front positions for all graphene coated and "As Received" samples were continuously monitored and recorded using a video camera. The transient permeability was estimated using Darcy's law [29]. Figure 8 presents in-plane permeability results based on flow front positions. The permeability decreases with increase in graphene content. This is mainly attributed to greater resistance offered by graphene inclusions blocking the channels for resin flow. Figure 9 shows optical microscopy images where the flow of graphene nanoparticles through the fiber bundles is evident. Figure 10 shows SEM micrographs of graphene coated glass composites at different magnifications. The resin flow path is shown out of the plane. The images show a good distribution of graphene nanoparticles in the composites structure. The circular cylinders are the glass fibers surrounded by irregular shaped grapheneplatelets. The nanoplatelets which are approximately 5 μm in diameter flow through the fiber bundles and tows of the reinforcement as the resin flow both in-plane and in transverse directions.

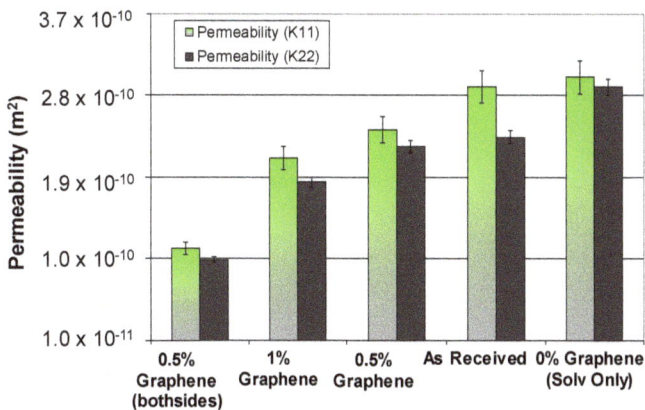

Figure 8. Transient permeability results of different percentages of graphene coated glass fiber reinforcement.

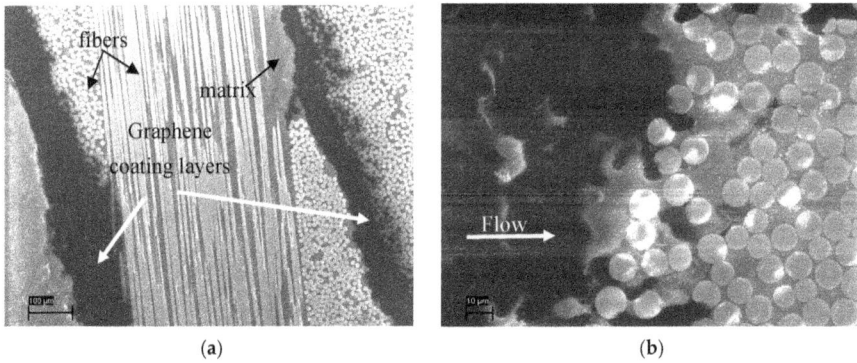

Figure 9. Optical microscopy images of graphene coated glass fiber composites at two different magnifications. (**a**) low magnification; (**b**) high magnification.

Figure 10. SEM images of graphene coated glass fiber composites at two different magnifications. (**a**) low magnification; (**b**) high magnification.

3.3. Mechanical Properties

3.3.1. Flexural Properties

Six specimens were tested for each of the samples. For all tests, the force vs. displacement graph was recorded and then stress and strain were calculated using the specimen dimensions. The stress strain plots for each set of data are shown in Figure 11. Overall, the curves were mostly linear with similar slopes before failure. The flexural modulus was calculated from the slope of stress strain curve, and the flexural strength was determined at the maximum stress.

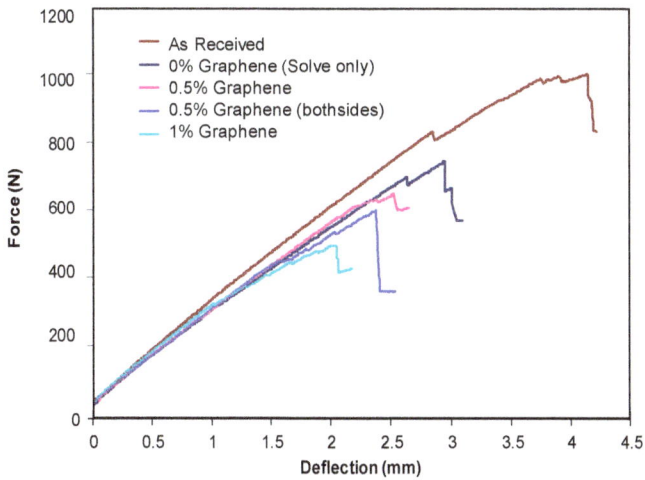

Figure 11. Load vs. deflection curves for different percentages of graphene coated glass fiber composites.

The plots in Figure 12a show the average flexural modulus and average flexural strength with error bars representing the range of the data for each set of experiments. The flexural modulus was highest with the 1% graphene amount while it remained almost constant for all other samples. The flexural strength decreased as the amount of graphene increased as shown in Figure 12b. The graphene used in this study was not functionalized, and it is anticipated that after treatment with resin compatible functional groups, the static and dynamic properties can be enhanced [12].

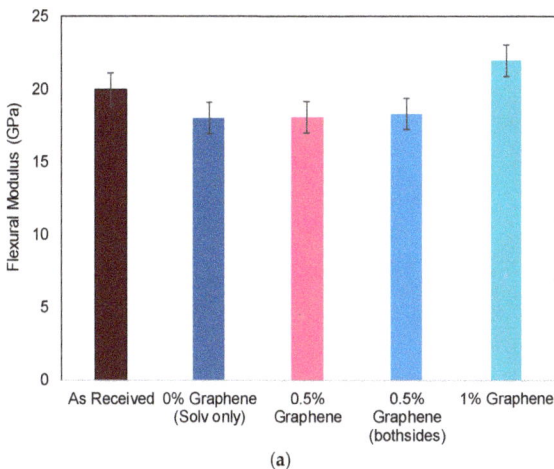

(a)

Figure 12. *Cont.*

(b)

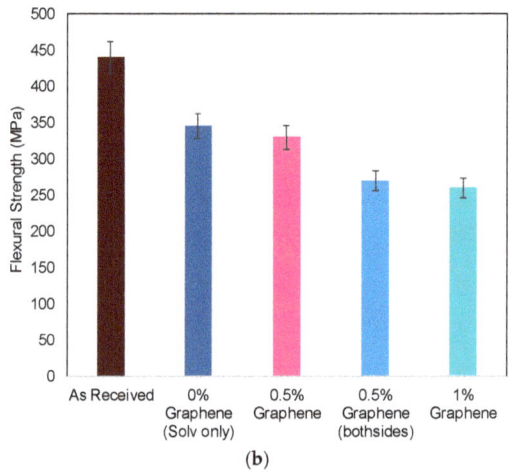

Figure 12. Flexural properties of different percentages of graphene coated glass fiber composites (**a**) flexural Modulus; (**b**) flexural Strength.

3.3.2. Low Velocity Impact

The data from the impact test system are presented in Figures 13 and 14. The figures represent plots of load vs. time and load vs. deflection of different configurations of samples at 20, 40, 60 and 80 J energies. The slope of the load-time curve, which is designated as the contact stiffness, increases with the increasing amount of energy. The initial knee found in the load-time plot is due to the inertia effect of the impactor and the sample. Once the inertia of the impactor and samples matched, a relatively smooth load rise is seen. Data collection is triggered by means of velocity detector just before the drop weight impacts the sample. In the current study, the impact phenomenon is characterized in terms of peak load and absorbed energy. Table 1 gives values of the peak load and absorbed energy of all types of samples impacted at four energy levels. The absorbed energy is calculated as the difference of total energy at the end of the event, and the energy at peak load. Energy absorption in composites is mainly through two modes: elastic strain energy and through various damage modes. The composite laminates are brittle in nature and respond elastically until they reach the peak load. If the impact energy is higher than the energy absorbed until the peak load, the additional energy is taken up in the creation of damage with a small amount of energy lost in friction between the sample and the impactor. As the impact energy is increased, the laminate undergoes large deformation. The next failure that takes place will be the tensile failure of the back surface due to flexure.

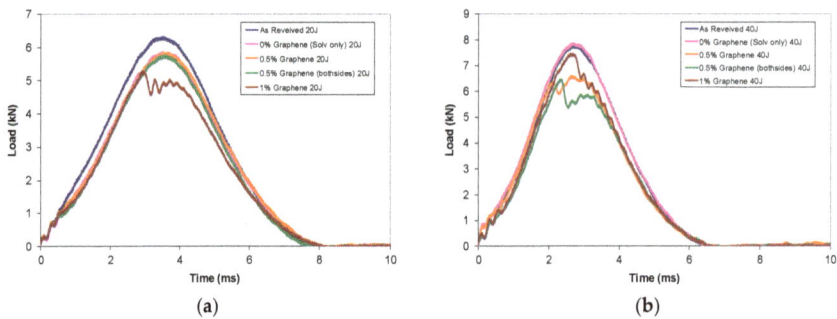

(a)

(b)

Figure 13. *Cont.*

(c)

(d)

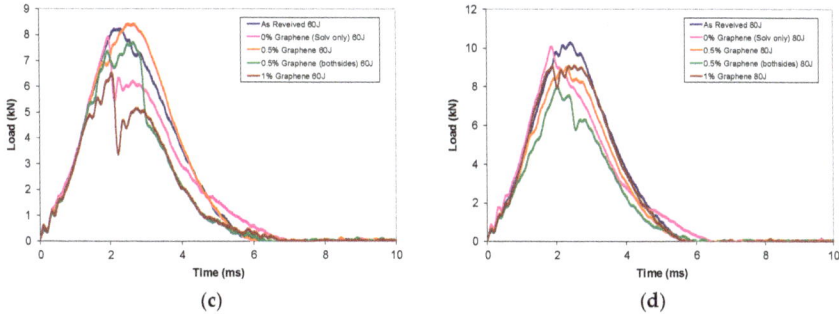

Figure 13. Load vs. time of glass fiber composites under impact (**a**) 20 J, (**b**) 40 J, (**c**) 60 J, and (**d**) 80 J.

Table 1. Peak load and absorbed energy.

	As Received	0% Graphene	0.5% Graphene	0.5% Graphene Bothsides	1% Graphene
20 J					
Peak Load (kN)	6.3	5.86	5.81	5.75	5.27
Absorbed energy (J)	0.1	0.49	0.14	1.44	5.75
40 J					
Peak Load (kN)	7.74	7.85	6.59	6.45	7.44
Absorbed energy (J)	12.05	11.71	13.57	18.7	14.21
60 J					
Peak Load (kN)	8.23	7.92	8.4	7.71	6.53
Absorbed energy (J)	26.48	29.86	21.1	14.33	24.93
80 J					
Peak Load (kN)	10.29	10.07	9.02	7.95	9.11
Absorbed energy (J)	24.12	39.02	29.1	33.55	23.79

In Figures 13 and 14, the sudden drop in force between 2 ms and 4 ms represents damage or delamination, mostly occurring in graphene coated samples. The damage in the "As Received" and 0% graphene is more localized and can be seen in the C-scan in Figure 15. The damage area was not clearly visible through naked eye for the graphene coated samples. The C-scan images showed that the damage area was in-between the plies due to delamination and this phenomena increased as the graphene concentration increased. It can be inferred that the inclusion of graphene into the samples resisted the propagation of cracks and through thickness penetration.

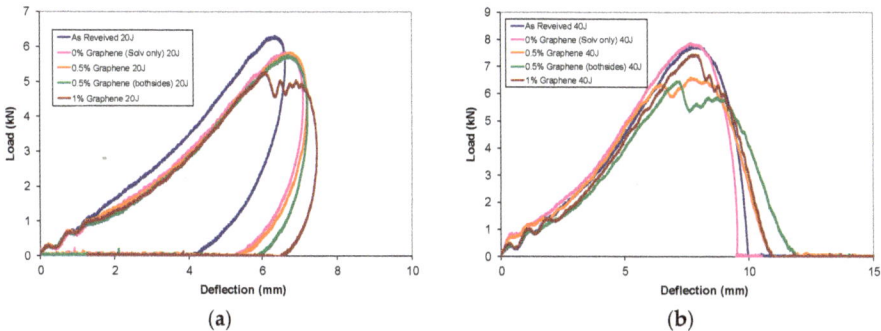

(a)

(b)

Figure 14. *Cont.*

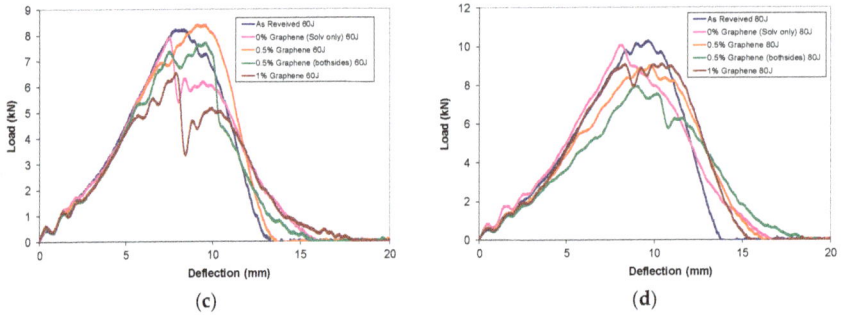

(c)

(d)

Figure 14. Load vs. deflection of glass fiber composites under impact (**a**) 20 J, (**b**) 40 J, (**c**) 60 J, and (**d**) 80 J.

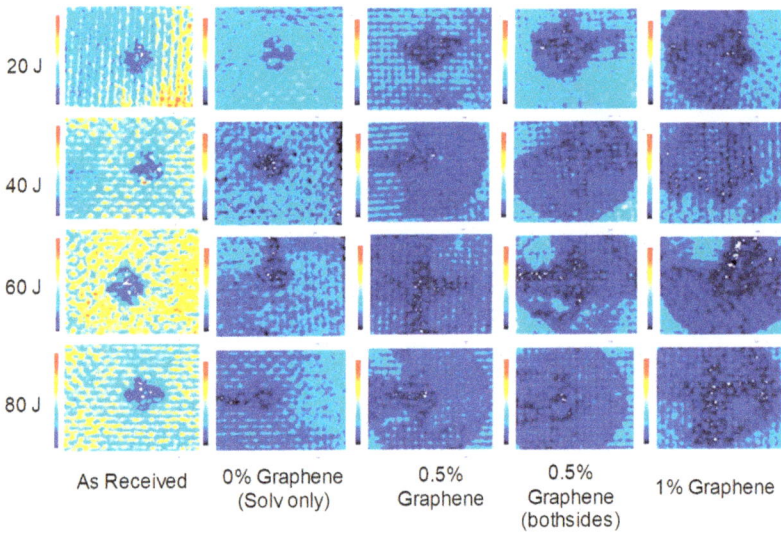

| 20 J | 40 J | 60 J | 80 J |

As Received | 0% Graphene (Solv only) | 0.5% Graphene | 0.5% Graphene (bothsides) | 1% Graphene

Figure 15. Ultrasonic images of impacted graphene coated glass fiber composites at different energy levels.

4. Conclusions

Graphene coating solutions were prepared by exfoliating graphite powder in a solvent using mechanical stirring and ultrasonication. Controllable amounts of graphene were successfully coated onto either one surface only or both surfaces of a glass fiber reinforcement using a brush coating technique. The solution coating was found viable method for introducing graphene in a reinforcing fabric for liquid composite molding process. The VARTM processing characteristics, such as compaction response and permeability of the coated fabrics were measured. The compaction response results show that as the graphene wt % increased, the resistance to compaction also increased. The transverse permeability of graphene coated samples did not change much except for the solvent treated samples, where the permeability was found to be higher due to possible removal or dissolution of the sizing in the solvent. The advancing front or transient permeability decreased with increase in graphene wt % which was mainly due to increase in resistance due to inclusions. Panels were manufactured for mechanical testing. The micrographs show that graphene dispersed well in the glass fabric with a very small amount of washout. The flexural modulus did not change much but the flexural strength decreased as the graphene wt % increased. The low velocity impact and C-scan

J. Compos. Sci. **2018**, *2*, 17

results show that the addition of graphene in a structure will absorb and distribute the energy under impact loading and resist through thickness damage.

Author Contributions: Rehan Umer conducted all the experiments and data analysis including writing of the manuscript.

Conflicts of Interest: The author declares no conflict of interest.

References

1. Potts, J.R.; Dreyer, D.R.; Bielawski, C.W.; Ruoff, R.S. Graphene-based polymer nanocomposites. *Polymer* **2011**, *52*, 5–25. [CrossRef]
2. Spitalsky, Z.; Tasis, D.; Papagelis, K.; Galiotis, C. Carbon nanotube-polymer composites: Chemistry, processing, mechanical and electrical properties. *Prog. Polym. Sci.* **2010**, *35*, 357–401. [CrossRef]
3. Kuilla, T.; Bhadra, S.; Yao, D.; Kim, N.H.; Bose, S.; Lee, J.H. Recent advances in graphene based polymer composites. *Prog. Polym. Sci.* **2010**, *35*, 1350–1375. [CrossRef]
4. Xu, Y.; Hong, W.; Bai, H.; Li, C.; Shi, G. Preparation and characterization of graphene/poly(vinyl alcohol) nanocomposites. *Carbon* **2009**, *47*, 3538–3543. [CrossRef]
5. Samad, Y.; Li, Y.; Alhassan, S.; Liao, K. Novel Graphene Foam Composite with Adjustable Sensitivity for Sensor Applications. *ACS Appl. Mater. Interfaces* **2015**, *7*, 9195–9202. [CrossRef] [PubMed]
6. Ponnamma, D.; Guo, Q.; Krupa, I.; Al-Maadeed, M.A.S.A.; Varughese, K.T.V.; Thomas, S.; Sadasivuni, K.K. Graphene and graphitic derivative filled polymer composites as potential sensors. *Phys. Chem. Chem. Phys.* **2015**, *17*, 3954–3981. [CrossRef] [PubMed]
7. Lee, X.; Yang, T.; Li, X.; Zhang, R.; Zhu, M.; Zhang, H.; Xie, D.; Wei, J.; Zhong, M.; Wang, K. Flexible graphene woven fabrics for touch sensing. *Appl. Phys. Lett.* **2013**, *102*, 163117. [CrossRef]
8. Aguilera-Servin, J.; Miao, T.; Bockrath, M. Nanoscale pressure sensors realized from suspended graphene membrane devices. *Appl. Phys. Lett.* **2015**, *106*, 083103. [CrossRef]
9. Zhu, S.; Ghatkesar, M.; Zhang, C.; Janssen, G. Graphene based piezoresistive pressure sensor. *Appl. Phys. Lett.* **2013**, *102*, 161904. [CrossRef]
10. Kim, S.; Drzal, L.T. Thermal stability and dynamic mechanical behavior of exfoliated graphite nanoplatelets–LLDPE composites. *Polym. Compos.* **2009**, *31*, 755–761. [CrossRef]
11. Jiang, X.; Drzal, L.T. Multifunctional high density polyethylene nanocomposites produced by incorporation of exfoliated graphite nanoplatelets 1: Morphology and mechanical properties". *Polym. Compos.* **2010**, *31*, 1091–1098. [CrossRef]
12. Biswas, S.; Fukushima, H.; Drzal, L.T. Mechanical and electrical property enhancement in exfoliated graphene nanoplatelet/liquid crystalline polymer nanocomposites. *Compos. Part A* **2011**, *42*, 371–375. [CrossRef]
13. Stankovich, S.; Dikin, D.A.; Dommett, G.H.B.; Kohlhaas, K.M.; Zimney, E.J.; Stach, E.A.; Piner, R.D.; Nguyen, S.T.; Ruoff, R.S. Graphene-based composite materials. *Nat. Lett.* **2006**, *442*, 282–286. [CrossRef] [PubMed]
14. Sun, L.; Gibson, R.F.; Gordaninejad, F.; Suhr, J. Energy absorption capability of nanocomposites: A review. *Compos. Sci. Technol.* **2009**, *69*, 2392–2409. [CrossRef]
15. Singh, V.; Joung, D.; Zhai, L.; Das, S.; Khondaker, S.I.; Seal, S. Graphene based materials: Past, present and future. *Prog. Mater. Sci.* **2011**, *56*, 1178–1271. [CrossRef]
16. Ávila, A.F.; Soares, M.I.; Neto, A.S. A Study on nanostructured laminated plates behavior under low-velocity impact loadings. *Int. J. Impact Eng.* **2007**, *34*, 28–41. [CrossRef]
17. Fan, Z.; Hsiao, K.T.; Advani, S.G. Experimental investigation of dispersion during flow of multi-walled carbon nanotube/polymer suspension in fibrous porous media. *Carbon* **2004**, *42*, 871–876. [CrossRef]
18. Lin, L.Y.; Lee, J.H.; Hong, C.E.; Yoo, G.H.; Advani, S.G. Preparation and characterization of layered silicate/glass fiber/epoxy hybrid nanocomposites via vacuum-assisted resin transfer molding (VARTM). *Compos. Sci. Technol.* **2006**, *66*, 2116–2125. [CrossRef]
19. Zhou, Y.; Pervin, F.; Rangari, V.K.; Jeelani, S. Fabrication and evaluation of carbon nano fiber filled carbon/epoxy composite. *Mater. Sci. Eng. A* **2006**, *426*, 221–228. [CrossRef]
20. Movva, S.; Zhou, G.; Guerra, D.; Lee, L.J. Effect of Carbon nano fibres on mold filling in a vacuum assisted resin transfer molding system. *J. Compos. Mater.* **2009**, *43*, 611–620. [CrossRef]

21. Peila, R.; Seferis, J.C.; Karaki, T.; Parker, G. Effects of nanoclay on the thermal and rheological properties of a VARTM Epoxy Resin. *J. Therm. Anal. Calorim.* **2009**, *96*, 587–592. [CrossRef]

22. Mahrholz, T.; Stängle, J.; Sinapius, M. Quantification of the reinforcement effect of silica nanoparticles in epoxy resins used in liquid composite molding processes. *Compos. Part A* **2009**, *40*, 235–243. [CrossRef]

23. Umer, R.; Li, Y.; Dong, Y.; Haroosh, H.J.; Liao, K. The effect of graphene oxide (GO) nanoparticles on the processing of epoxy/glass fiber composites using resin infusion. *Int. J. Adv. Manuf. Technol.* **2015**, *81*, 2183–2192. [CrossRef]

24. Ali, M.A.; Umer, R.; Khan, K.A.; Samad, Y.A.; Liao, K.; Cantwell, W.J. Graphene coated piezo-resistive fabrics for liquid composite molding process monitoring. *Compos. Sci. Technol.* **2017**, *148*, 106–114. [CrossRef]

25. Wu, H.; Rook, B.; Drzal, L.T. Dispersion optimization of exfoliated graphite nanoplatelets in polyetherimide nanocomposites: Extrusion vs. precoating vs. solid state ball milling. *Polym. Compos.* **2009**, *34*, 426–432. [CrossRef]

26. Maenosono, S.; Okubo, T.; Yamaguchi, Y. Overview of nanoparticle array formation by wet coating. *J. Nanopart. Res.* **2003**, *5*, 5–15. [CrossRef]

27. Rachmadini, Y.; Tan, V.B.C.; Tay, T.E. Enhancement of Mechanical Properties of Composites through Incorporation of CNT in VARTM-A Review. *J. Reinf. Plast. Compos.* **2010**, *29*, 2782–2807. [CrossRef]

28. Yavari, F.; Rafiee, M.A.; Rafiee, J.; Yu, Z.Z.; Koratkar, N. Dramatic increase in fatigue life in hierarchical graphene composites. *ACS Appl. Mater. Interfaces* **2010**, *2*, 2738–2743. [CrossRef] [PubMed]

29. Simmons, C.T.; Darcy, H. Immortalised by his scientific legacy. *Hydrogeol. J.* **2008**, *16*, 1023–1038. [CrossRef]

© 2018 by the author. Licensee MDPI, Basel, Switzerland. This article is an open access article distributed under the terms and conditions of the Creative Commons Attribution (CC BY) license (http://creativecommons.org/licenses/by/4.0/).

Journal of
composites science

MDPI

Article

Preparation and Performance of Ecofriendly Epoxy/Multilayer Graphene Oxide Composites with Flame-Retardant Functional Groups

Ming-He Chen, Cing-Yu Ke and Chin-Lung Chiang *

Green Flame Retardant Material Research Laboratory, Department of Safety, Health and Environmental Engineering, Hung-Kuang University, Taichung 433, Taiwan; n74731@gmail.com (M.-H.C.); h22432003@gmail.com (C.-Y.K.)
* Correspondence: dragon@sunrise.hk.edu.tw

Received: 23 January 2018; Accepted: 21 March 2018; Published: 23 March 2018

Abstract: This study aimed to prepare ecofriendly flame retardants. Using the –OH and –COOH functional groups of multilayer graphene oxide (GO) for the hydrolytic condensation of tetraethoxysilane (TEOS), TEOS was grafted onto GO to form Si-GO. Subsequently, p-aminophenol (AP) was grafted onto Si-GO to produce Si-GA, forming composite materials with epoxy (EP). The structures and properties of the composite materials were examined with Fourier-transform infrared spectroscopy (FTIR), thermogravimetric analysis (TGA), and the limiting oxygen index (LOI). In terms of structure, FTIR observed two characteristic peaks of Si-GO, namely Si–O–C and Si-O-Si, indicating that TEOS was successfully grafted onto GO. TGA was used to determine the thermal stability of the epoxy/Si-GA composites; with the increase in Si-GA, the char yield of the materials increased from 15.6 wt % (pure epoxy) to 25 wt % (epoxy/10 wt % Si-GA), indicating that Si-GA effectively enhanced the thermal stability of the epoxy matrix. Lastly, the flame retardant tests determined that the LOI value rose from 19% (pure epoxy) to 26% (epoxy/10 wt % Si-GA), proving that graphene with modified silicon can be used to enhance the flame retardancy of epoxy.

Keywords: epoxy; graphene oxide; silicon; composite; flame retardant

---

1. Introduction

Polymer materials have a wide range of applications. However, their popularity has also necessitated the development of their flame resistance, either by increasing the flame retardancy of a material itself, or through the addition of flame retardants. Epoxy is a type of thermosetting polymer that has been widely applied in numerous fields including adhesives, coatings, package-on-package, electrical insulation, and composite materials [1,2]. However, although epoxies are high performance materials with excellent mechanical and chemical properties, they are also highly combustible. This has actuated the demand for flame-proof epoxy, of which the most widely researched are epoxies that are both incombustible and halogen-free in accordance with the WEEE (Waste Electrical and Electronic Equipment) and RoHS (Restriction of Hazardous Substances) directives [3–5]. In recent years, the call for "halogen free" products, initiated by the European Union, has had a considerable effect on the plastics industry, because more than 50% of their products use flame retardants containing halogen. The demands for halogen-free flame retardants, coupled with the used of nanotechnologies, will indirectly lead to improvements in material incombustibility [6,7]. This indicates the immense business opportunities behind such successful research, which has triggered fierce competition worldwide regarding related research.

Graphene is one kind of nanomaterial with distinctive physical, chemical, and mechanical properties, such as high electron mobility, high mechanical strength, and high thermal conductivity, which makes graphene one of the most favorable choices for fabricating composite materials with polymers matrix [8,9]. As the addition of graphene is known to substantially improve the mechanical, electrical, and thermal properties of the material, research has proved that adding even miniscule amounts of graphene can effectively increase the flame retardant property of materials [10,11]. The agglomeration of graphene sheets due to the strong van der Waals forces among their sheets and the weak compatibility with most of polymer matrices had been fundamental roadblocks that restricted its potential as a reinforcing agent [12]. Surface modification of graphene by adding functional groups is an effective way to reduce the tendency to agglomerate [13]. In addition, functionalization increases the graphene compatibility with specific polymers improving the reinforcing effect [14]. Graphene was grafted flame retardant components or functional groups on its surfaces to possess better thermal stability and flame retardant properties [15–18]. In our study, we also proved that functionalized graphene nanosheets were well dispersed in organic solvents, meaning that the modified fillers will be uniformly dispersed in polymer matrix [19].

In the present study, varying percentages of modified graphene oxide (GO) were added to epoxy matrix, which then underwent Fourier-transform infrared spectroscopy (FTIR), thermogravimetric analysis (TGA), and limiting oxygen index (LOI) to determine their properties and render them useful for future applications.

2. Experimental

2.1. Materials

Multilayer graphene nanosheets (Knano Graphene Technology Corporation Limited, Xiamen, China) were used as the starting material for the preparation of graphene nanosheets oxide. The graphene nanosheets had a thickness of 5–15 nm, a purity of 99.5%, and a density of 2.25 g/cm^3. Diglycidyl ether of bisphenol-A (DGEBA type) was provided from Chang Chun Petrochemical Co., Ltd., Miali, Taiwan. 4,4′-diaminodiphenylmethane (DDM) was purchased from Acros Chemical Co., Mullica Hill, NJ, USA. Sulfuric acid (95.7 wt %), phosphorous acid (85 wt %), hydrogen chloride (37 wt %), ethanol (95 wt %), hydrogen peroxide (31 wt %) and potassium permanganate were purchased from Echo Chemical Co. Ltd., Miali, Taiwan. Tetraethyl silicate (TEOS) was purchased from Sigma-Aldrich Co. St. Louis, MI, USA. p-Aminophenol (AP) was purchased from Alfa Aesar Co., Ward Hill, MA, USA.

2.2. Preparation of GO

Four 400 mL serum vials, each containing 3 g of multilayer graphene nanosheets and filled with 400 mL of ethanol, were ultrasonicated for 4 h (water level > 400 mL, water changed every 1 h). After suction filtration, the contents were evenly dispersed on a rectangular aluminum plate, placed in an oven, and heated overnight at 80 °C. In an ice bath inside a fume hood, 3 g of multilayer graphene nanosheets was added into a mixture of sulfuric acid and phosphoric acid at a 9:1 volume ratio (360 mL:40 mL), followed by 18 g of potassium permanganate as a strong oxidant, and stirred for 12 h under ice bath. Subsequently, this was slowly added with 400 mL of deionized (DI) water in an ice bath against the fuming and exothermic reaction, and then 3 mL of 35% hydrogen peroxide was added dropwise. The mixture subsequently underwent centrifugal filtration at 6000 rpm, and was washed with DI water for 20 min several times. The filtered GO was then placed in an ice bath and added in sequence with 200 mL of water, 200 mL of hydrochloric acid, and 200 mL of ethanol (600 mL in total), stirred for 1 h, centrifugally filtered at 6000 rpm, and washed with ethanol for 20 min several times. Finally, the filtered material was washed with 200 mL of ethanol under suction filtration several times, and heated in a vacuum oven overnight at 80 °C; thus, the preparation of GO was completed.

J. Compos. Sci. **2018**, 2, 18

2.3. Preparation of Si-GO

GO (0.4 g) was added into 150 mL of an ethanol–DI water mixture (volume ratio = 5:1), which then underwent ultrasonication for 30 min. Subsequently, 5.6 mL of ammonia was added as a catalyst and stirred to form a stable and evenly mixed suspension. Subsequently, 1.5 mL of TEOS was added to the suspension, which then underwent ultrasonication for 30 min, and was left at room temperature for 15 h before receiving centrifugal filtration at 6000 rpm. The filtered substance was then washed with ethanol several times and dried overnight in a vacuum oven at 60 °C. Thus; the preparation of Si-GO was completed. The reaction equation is as shown in Scheme 1:

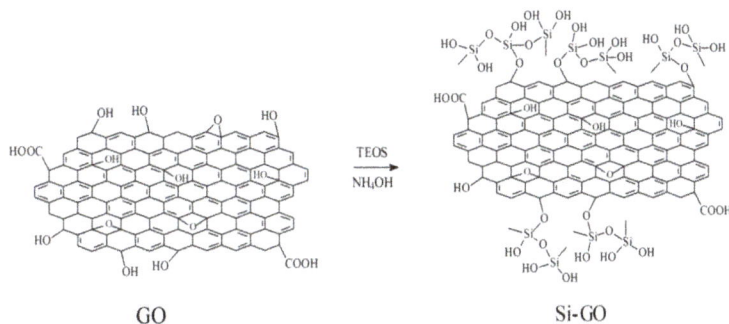

GO   Si-GO

Scheme 1. The reaction between GO and TEOS.

2.4. Preparation of the Epoxy/Si-GA Composite

In a serum vial, 0.3 g of Si-GO was mixed with 1 g of AP, and mechanically mixed and stirred with ethanol at 50 °C for 3 h to form Si-GA. Specimens of Si-GA at 1 wt %, 5 wt %, and 10 wt % were then mixed into 10 g epoxy, and mechanically stirred with 1.38 g of DDM (weight ratio = DDM/epoxy = 0.138) as the curing agent. Subsequently, each of the mixed specimens was poured onto an aluminum plate, and placed in an oven to be heated for 2 h at 180 °C; the resulting material was a cured epoxy/Si-GA composite. The reaction scheme was shown in Scheme 2. Epoxy/Si-GO composite was prepared by the same process without AP, which was used as the comparison.

2.5. Measurements

The FTIR spectra of the materials were obtained between 4000–400 cm^{-1} using a Nicolet Avatar 320 FT-IR spectrometer from ALT (Alexandria, VA, USA). Thin films were prepared by solution-casting. A minimum of 32 scans were signal-averaged with a resolution of 2 cm^{-1}.

X-ray photoelectron spectra (XPS) were recorded using a PHI Quantera SXM/Auger (Ulvac-Phi. Inc., Chigasaki, Kanagawa, Japan), with Al Ka excitation radiation (hv = 1486.6 eV). The pressure in the analyzer was maintained at around 6.7×10^{-7} Pa. XPS data was processed using a DS 300 data system (Ulvac-Phi. Inc., Chigasaki, Kanagawa, Japan).

The thermal degradation of the composites was examined using a thermogravimetric analyzer (TGA) (Perkin Elmer TGA 7, Waltham, MA, USA) from room temperature to 800 °C at a heating rate of 10 °C/min in an atmosphere of nitrogen. The measurements were made on 6–10 mg samples.

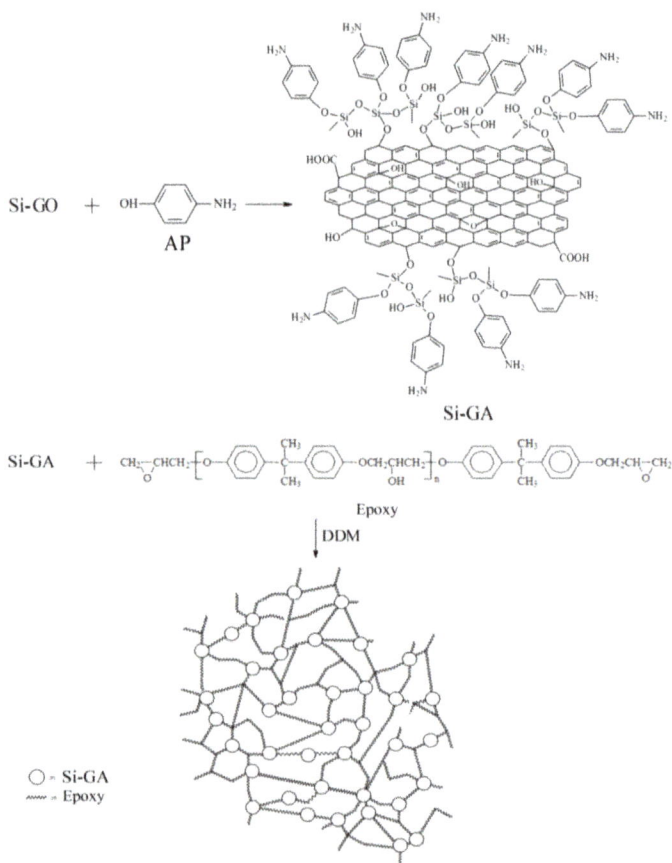

Scheme 2. The preparation of epoxy/Si-GA composites.

The LOI test was carried out following the ASTM (American Society for Testing and Materials) D 2863 Oxygen Index Method. The epoxy/Si-GA composites were put into the oven for 1 and 5 min at 600 °C, respectively. The resulting chars of the samples were tested by Raman spectra. Raman spectra were recorded using a Lab Ram I confocal Raman spectrometer (Montpellier Cedex, France). An He-Ne laser with a laser power of about 15 mW at the sample surface was utilized to provide an excitation wavelength of 632.8 nm. A holographic notch filter reflected the exciting line into an Olympus BX40 microscope, Olympus, Tokyo, Japan.

3. Results and Discussion

3.1. Characterization of Grafting Reaction

FTIR was employed to determine the structure of the specimens, and monitor the changes of the functional group during the reaction. Figure 1 shows the results for GO and Si-GO, and indicates that the characteristic absorption peaks of Si-GO functional groups were observed at 1120 cm^{-1} (Si–O–C), 1039 cm^{-1} (Si–O–Si), and 453 cm^{-1} (Si–O–Si) [20–23]. The presence of the Si–O–Si and Si–O–C functional groups in the infrared spectrum of Si-GO suggests that TEOS had been grafted onto GO.

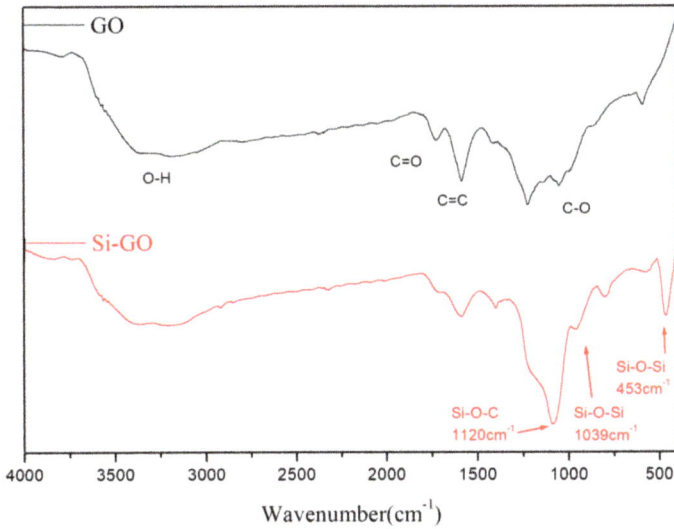

Figure 1. FTIR spectra of GO, Si-GO.

X-ray photoelectron spectroscopy (XPS) was employed to determine the surface compositions of GO and Si-GO. Figures 2 and 3 are the XPS C1s spectra of GO and Si-GO, respectively. Similarly, Figures 4 and 5 are the XPS O1s spectra of GO and Si-GO, respectively.

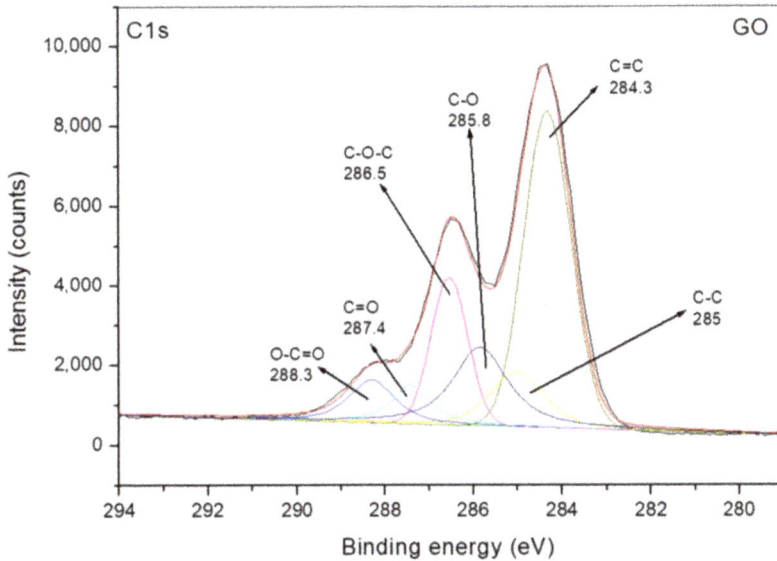

Figure 2. C1s XPS spectra of GO.

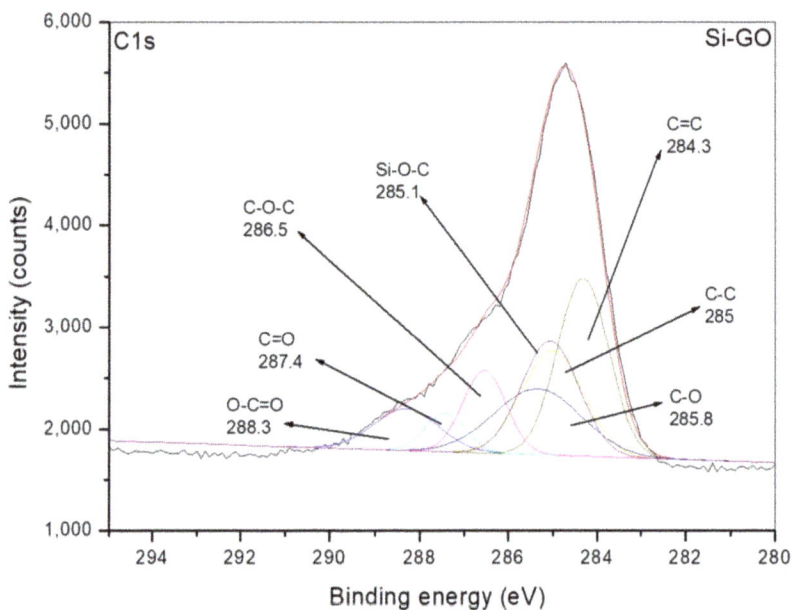

Figure 3. C1s XPS spectra of Si-GO.

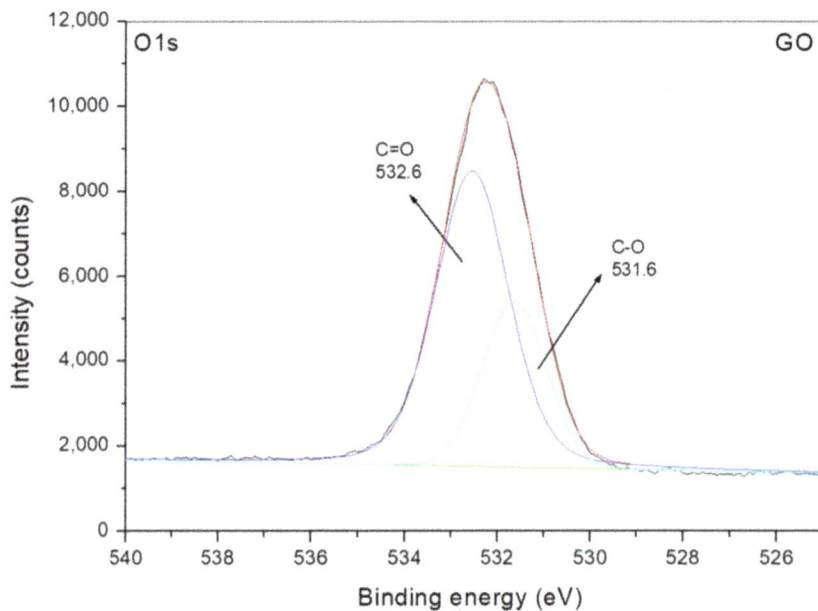

Figure 4. O1s XPS spectra of GO.

In Figure 2 of C1s spectra of GO, the characteristic peaks of C–C and C=C appear at 285 eV and 284.3 eV, respectively. Moreover, oxygen-carrying functional groups of O–C=O, C=O, C–O–C, and C–O are observed at 288.3 eV, 287.4 eV, 286.5 eV, and 285.8 eV, respectively. This proves that multilayer graphene nanosheets had been successfully oxidized [24,25]. In Figure 3 for C1s spectra of Si-GO, a new peak of Si–O–C appears at 285.1 eV, indicating that the silicon-carrying TEOS had been grafted onto the surface of GO through a condensation reaction. Furthermore, the facts that the peaks of C–O and C=O are observed at 532.6 eV and 531.6 eV, respectively, in Figure 4 for O1s spectra of GO, and peaks of Si–O–C and Si–O–Si are observed at 533.5 eV and 533 eV [25], respectively, in Figure 5 for O1s spectra of Si-GO, verifies that the silicon-and oxygen-carrying TEOS had been successfully grafted onto GO.

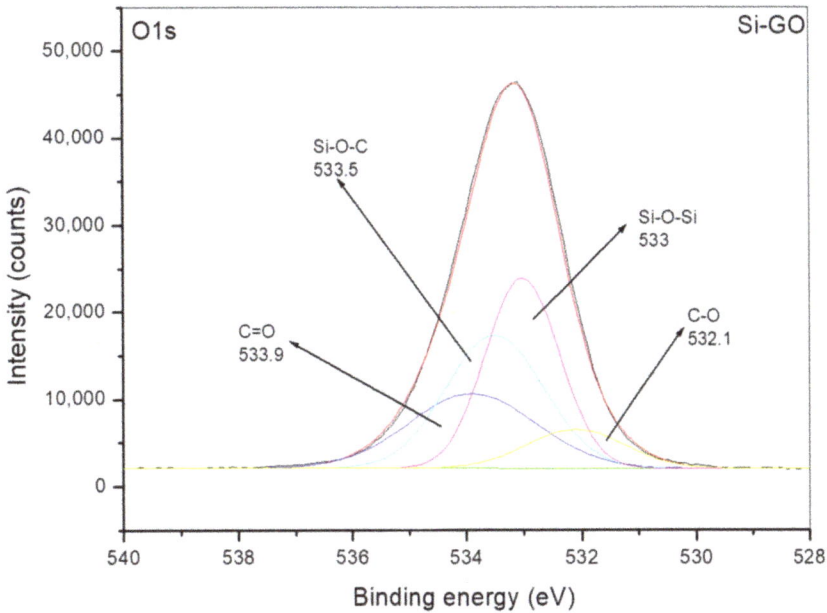

Figure 5. O1s XPS spectra of Si-GO.

3.2. Thermal Properties of Epoxy/Si-GA Composites

TGA was employed to determine the thermal stability through thermogravimetric weight loss–temperature curves of the specimens at 800 °C. Figure 6 shows the TGA results of the thermal stability test of the composite materials for Si-GO, pure epoxy, 10 wt % of epoxy/GO, and 10 wt % of epoxy/Si-GA under a nitrogen-filled environment and 20 °C/min heating rate.

There were two steps of thermal degradation for Si-GO from Figure 6. At temperatures below 200 °C, it may be caused by the physical adsorption of moisture on the carbon plane or by the hydrogen bonding of water molecules with the carbon planar structure. Above 200 °C, it could be the dehydration of hydroxyl functional groups of TEOS, and it would lead to the loss of the weight. At 800 °C, the char yield of Si-GO was 54.9 wt %. Because graphene has a relatively large surface area, it created more char and helped improve the thermal stability of the materials. The char yield of epoxy/Si-GA (10 wt %) was 25.0 wt %, which was higher than the 20.4 wt % of epoxy/GO (10 wt %). The addition of AP into epoxy matrix could improve the dispersion of Si-GO in the composite. AP was used as a coupling agent to connect epoxy with Si-GO together. The combination of SiO_2 and GO further enhanced the thermal stability of Si-GA because a layer of SiO_2 was formed on the surface of GO, which prevented thermal degradation from occurring. This verified that the grafting of silicon-containing structures onto GO effectively improved the thermal stability of GO composite materials.

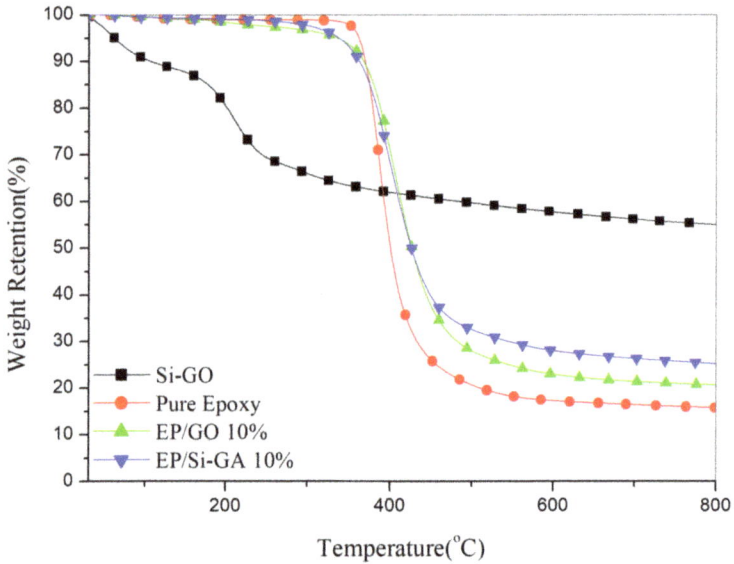

Figure 6. Comparison of calculated and experimental TGA curves for EP/Si-GA 10%.

Figure 7 indicates that adding 1 wt %, 5 wt %, and 10 wt % of Si-GA into epoxy gradually raised the char yield from 15.6 wt % (pure epoxy) to 25.0 wt % (10 wt % of epoxy/Si-GA). This suggests that the addition of flame retardant can effectively increase the thermal stability of a composite, and the higher the percentage of flame retardant, the greater the thermal stability. Figure 8 shows that, at 350–450 °C, the maximum thermal degradation rate of 10 wt % of epoxy/Si-GA was −15.1 wt %/min, which was noticeably smaller than the −31.6 wt %/min for epoxy matrix, again verifying that the addition of flame retardant effectively delayed the thermal degradation of epoxy.

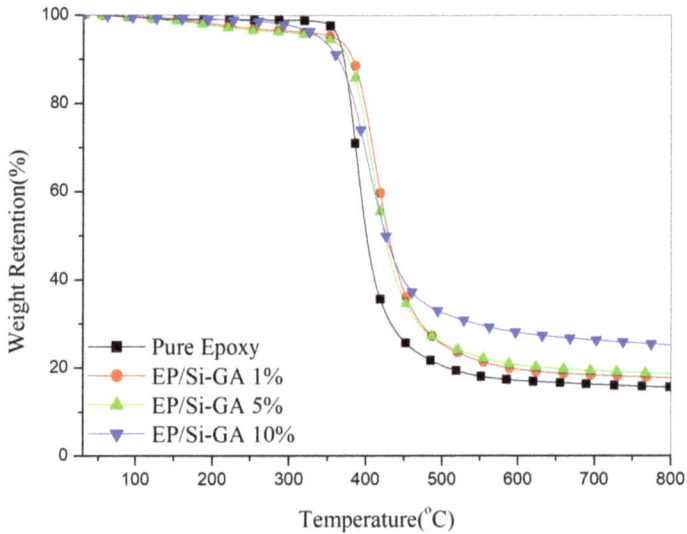

Figure 7. Thermogravimetric curves of epoxy with various Si-GA contents.

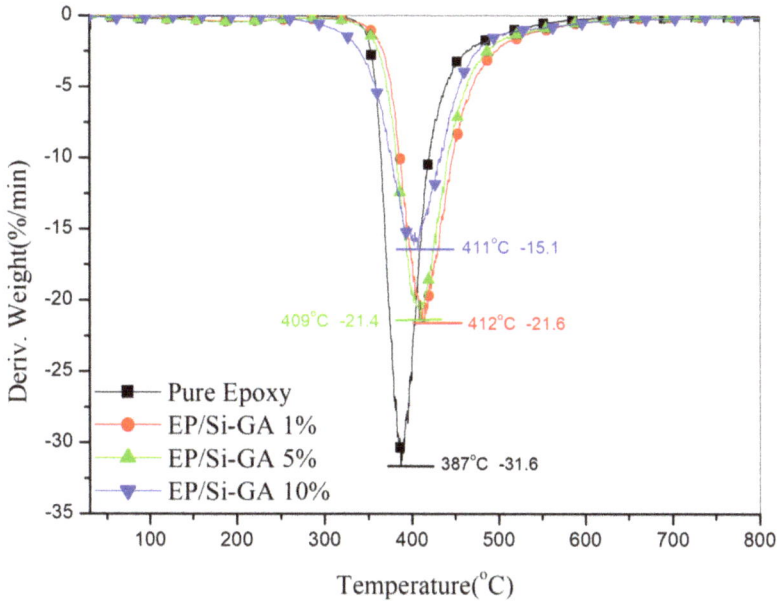

Figure 8. Derivative curves of epoxy with various Si-GA contents.

3.3. Interaction between Organic and Inorganic Phases

By comparing the theoretical values and the experimental values in the TGA curves, this study verified the thermal degradation behaviors of the composite materials [26]. Taking epoxy/Si-GA composites as an example, the comparison of its calculated theoretical values (pure epoxy × 0.9 + Si-GA × 0.1) with the experimental TGA curve of epoxy/Si-GA 10% is shown in Figure 9. The theoretical value was calculated from simple linear combination of the values of pure epoxy and Si-GA by TGA, meaning that no interaction occurs between organic and inorganic phases. The experimental value was indicated as larger compared with the theoretical curve, suggesting an improved interaction forces between organic and inorganic phases. This observation verified that, when Si-GA was added into epoxy as a flame retardant, the enhancements of interaction forces between the two phases effectively improved the thermal stability of the composites. The TGA curves also proved that the thermal stability was even greater than theoretically expected.

3.4. Thermal Stability of the Composites

This analysis relied on the integrated procedure decomposition temperature (IPDT), which can be obtained from the TGA data, as the indicator of thermal stability for composite materials [27]. Specifically, integration was used to calculate the area under the thermal degradation curve as shown in Figure 10 [28,29], which was then substituted into the following formula for the material's IPDT. A higher IPDT indicated greater thermal stability of the composites. In Table 1, the IPDT values increased with the concentration of Si-GA, which increased from 629 °C for pure epoxy to 825 °C for 10 wt % of epoxy/Si-GA; this was a 196 °C increase. In summary, the higher a material's initial thermal degradation temperature and char yield, the greater its thermal stability, and the higher the IPDT value, the greater the thermal stability.

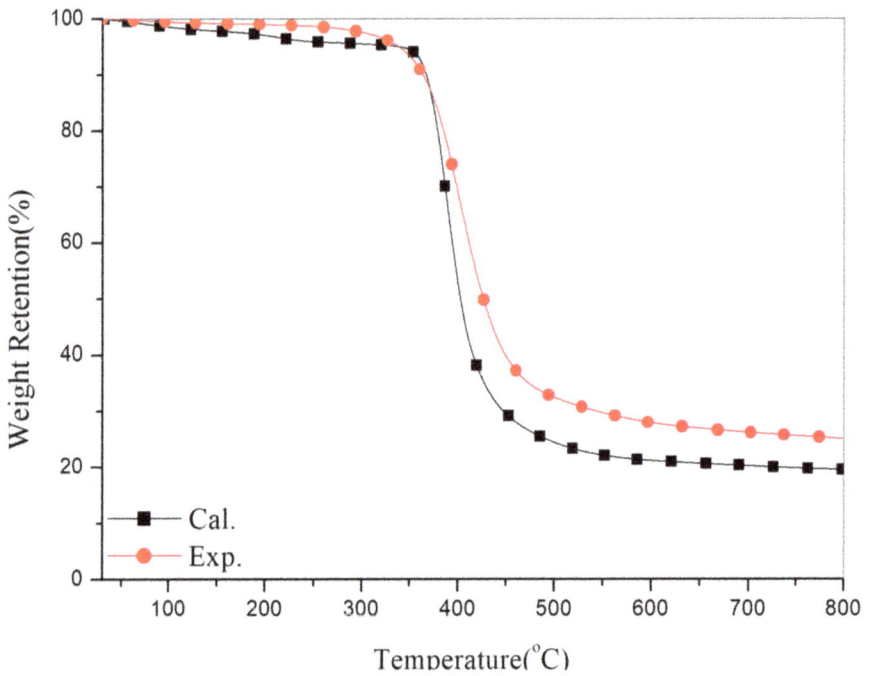

Figure 9. Comparison of calculated and experimental TGA curves for EP/Si-GA 10%.

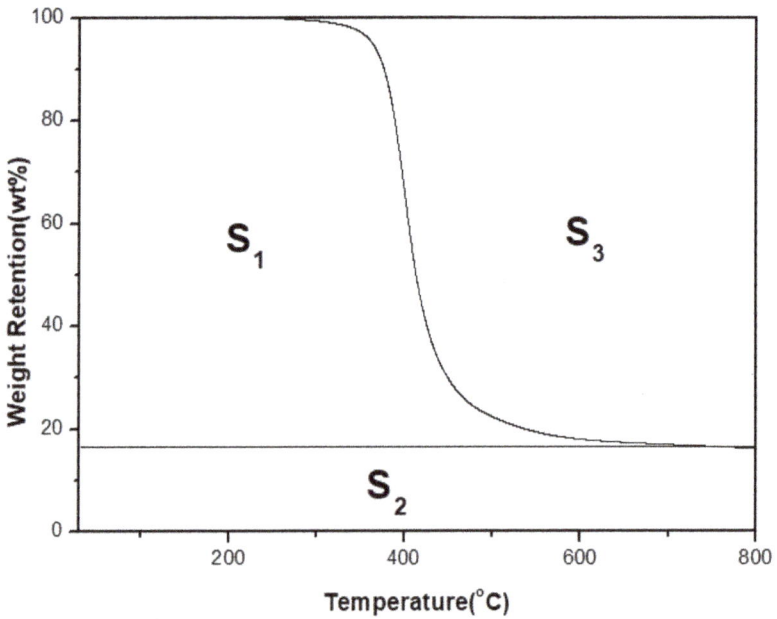

Figure 10. TGA curve of the materials.

Table 1. Thermal properties of epoxy with various contents Si-GA.

Sample No.	T_{max} (°C)	R_{max} (wt %/min)	IPDT[1] (°C)	C.Y.[2] (wt %)
Pure Epoxy	387	−31.6	629	15.6
Epoxy/Si-GA 1%	409	−21.6	672	17.6
Epoxy/Si-GA 5%	412	−21.4	689	18.3
Epoxy/Si-GA 10%	411	−15.6	825	25.0

[1] IPDT means integrated procedure decomposition temperature. [2] C.Y. means char yield.

The formula is as follows:

$$\text{IPDT (°C)} = A^* \times K^* \times (Tf - Ti) + Ti,$$

where

Ti = the initial experimental temperature (100 °C),
Tf = the final experimental temperature (800 °C),
$A^* = (S1 + S2)/(S1 + S2 + S3)$,
$K^* = (S1 + S2)/S1$.

Substituting the Ti, Tf, S1, S2, and S3 values into the formula yields the IPDT value.

3.5. Kinetics of Thermal Degradation

For materials to initiate thermal degradation, they must overcome an energy barrier, which is the activation energy of the materials. In this study, the TGA results of varying heating rates (5 °C/min, 10 °C/min, 20 °C/min, and 40 °C/min) under the nitrogen-filled environment were used to determine the thermal degradation kinetic characters of epoxy/Si-GA composites, and analyze the changes in their activation energies by Ozawa's method [30].

According to Table 2, the activation energies (ΔE) of pure epoxy and epoxy/Si-GA 10 wt % were 130.2 kJ/mole and 180.2 kJ/mole, respectively. This suggests an increase in the activation energy of the composite material after the addition of flame retardant. High activation energy of a composite material indicates that it is less prone to thermal degradation, or, in other words, has a high thermal stability. The results of the present study were in accordance with the conclusion drawn from the IPDT analysis.

Table 2. The calculated activation energy of thermal degradation with various conversions by Ozawa method.

α	Pure Epoxy		Epoxy/Si-GA 10%	
	E(kJ/mole)	R Value	E (kJ/mole)	R Value
0.2	146.8	0.88	212.8	0.99
0.3	138.9	0.92	194.3	0.99
0.4	131.6	0.94	179.2	0.99
0.5	126.8	0.95	173.6	0.99
0.6	122.8	0.96	169.4	0.99
0.7	121.8	0.96	170.0	0.99
0.8	123.5	0.94	162.0	0.98
ΔE(av)	130.3		180.2	

α = 0.2~0.8.

3.6. Flame Retardancy of the Materials

Limiting oxygen index (LOI) was used to determine the flammability of composite materials with varying concentrations. The flammability of materials was determined by controlling the nitrogen and oxygen concentrations in a confined space in accordance with the LOI results. Because the concentration of oxygen in the atmosphere is approximately 21%, flame retardant materials that require an oxygen content of more than 21% are difficult to burn in the natural environment. The concentrations of oxygen in confined space can be used as the index of flammability for the materials according to following equation. Therefore, the combustibility of the materials was set: materials with an LOI < 21% were classified as combustible and LOI > 21% as self-extinguishing [31,32]. Furthermore, the flow rates $[O_2]$ and $[N_2]$ were set using the following equation (mL/s):

$$\text{LOI (\%)} = \frac{[O_2]}{[O_2] + [N_2]} \times 100. \tag{1}$$

According to Figure 11, pure epoxy, 10 wt % epoxy/GO composite, and 10 wt % of epoxy/Si-GA were types of highly combustible and self-extinguishing polymers, respectively (LOI of 19%, 23%, and 26%). In general, the LOI values ascended, indicating that the increase in modified Si-GO also increased the flame retardancy of epoxy.

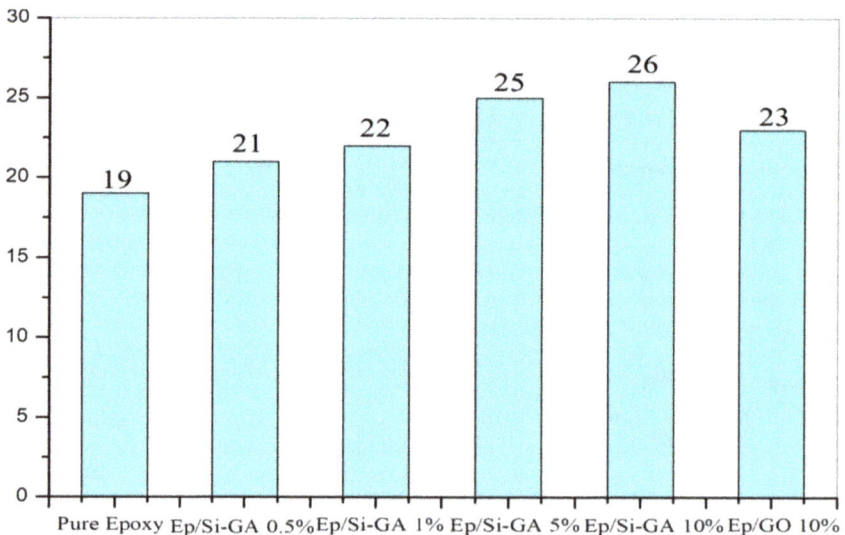

Figure 11. The flame retardant property of Epoxy/GO and Epoxy/Si-GA composites by LOI values.

3.7. Analysis of Char

Raman spectra was used to measure the 1-min and 5-min chars of epoxy/GO and epoxy/Si-GA composites at 600 °C, in order to observe the changes in their D band and G band shown as Figures 12 and 13. The D band (disordered band) occurred at 1350 cm^{-1}, and showed the sp^3 structure formed by C–C; by contrast, the G band (graphitic band) occurred at 1580 cm^{-1}, and showed the sp^2 structure formed by C=C [33,34]. The results are as listed in Table 3, which indicate that the D/G value of epoxy/GO burned for 5 min (0.42) is smaller than for that burned for 1 min (0.83), suggesting that the addition of GO increased the char formation of composite materials as they burned. Similarly, the D/G value of epoxy/Si-GA that has burned for 5 min (0.04) is smaller than that burned for 5 min (0.41). The change in the D/G values of the two materials indicates that silicon facilitated graphitic char formation, suggesting that the addition of Si-GA enabled the formation of a graphitic char layer during combustion and thus improved the material's thermal stability.

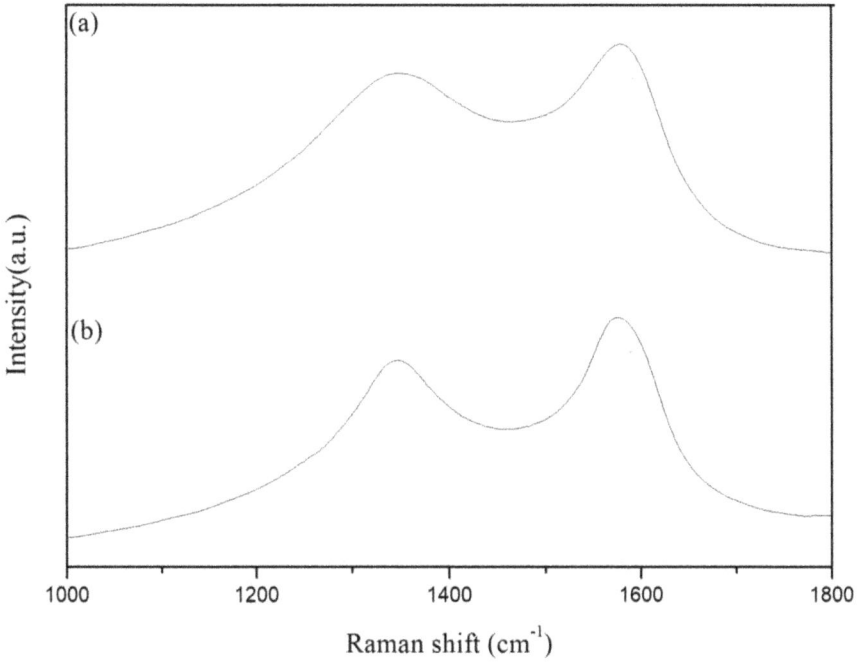

Figure 12. Raman spectra of char products from EP/GO 10 at 600 °C (**a**) 1 min; (**b**) 5 min.

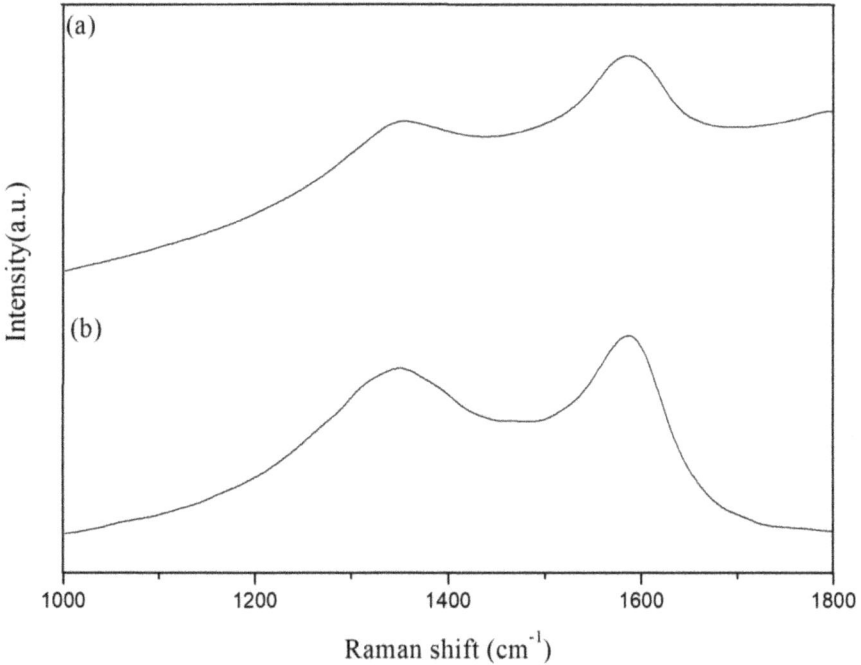

Figure 13. Raman spectra of char products from EP/Si-GA 10 at 600 °C (**a**) 1 min; (**b**) 5 min.

Table 3. The area ratio of Raman shift for Epoxy/GO & Epoxy/Si-GA composites char.

Sample No.		D-Band	G-Band	D/G
		1356 cm^{-1}	1580 cm^{-1}	
Epoxy/GO 10	1 min	679,856	816,332	0.83
	5 min	323,024	768,821	0.42
Epoxy/Si-GA 10	1 min	3.97×10^6	6.50×10^6	0.61
	5 min	302,028	7.60×10^6	0.04

4. Conclusions

This study successfully used the sol–gel method to prepare Si-GO, which was then used to fabricate composite materials with epoxy. Subsequently, FTIR and XPS analyses proved that TEOS was grafted onto GO. The IPDT values increased with the concentration of Si-GA, which increased from 629 °C for pure epoxy to 825 °C for 10 wt % of epoxy/Si-GA; this was a 196 °C increase. The results of TGA, DTG, and IPDT all verified that the addition of Si-GA effectively enhanced the thermal stability of the base material. This was further verified by the activation energy values of the materials; pure epoxy and epoxy/Si-GA 10 wt % were 130.2 kJ/mole and 180.2 kJ/mole, respectively, a 38.46% increase, which suggested that adding silicon-carrying fire retardant raised the activation energy required for thermal degradation and compelled thermal degradation to overcome an even higher energy threshold. Pure epoxy and 10 wt % of epoxy/Si-GA were highly combustible and flame retardant polymers, respectively (LOI of 19% and 26%). Results of the LOI analysis indicated that the epoxy composite materials achieve high flame retardancy and can be classified as flame-retardant. Raman spectra determined that the forming of graphite-based char enhanced the materials' thermal stability and incombustibility.

Acknowledgments: The authors would like to express their appreciation to the Ministry of Science and Technology of the Republic of China for financial support of this study under grant MOST-105-2221-E-241-001-MY3.

Author Contributions: Ming-He Chen and Chin-Lung Chiang conceived and designed the experiments; Ming-He Chen performed the experiments; Cing-Yu Ke and Chin-Lung Chiang analyzed the data; Chin-Lung Chiang wrote the paper.

Conflicts of Interest: The authors declare no conflict of interest.

References

1. Pan, G.; Du, Z.; Zhang, C.; Li, C.; Yang, X.; Li, H. Synthesis, characterization, and properties of novel novolac epoxy resin containing naphthalene moiety. *Polymer* **2007**, *48*, 3686–3693. [CrossRef]
2. Akatsuka, M.; Takezawa, Y.; Amagi, S. Influences of inorganic fillers on curing reactions of epoxy resins initiated with a boron trifluoride amine complex. *Polymer* **2001**, *42*, 3003–3007. [CrossRef]
3. Perret, B.; Schartel, B.; Stös, K.; Ciesielski, M.; Diederichs, J.; Döring, M.; Krämer, J.; Altstädt, V. Novel DOPO-based flame retardants in high-performance carbon fiber epoxy composites for aviation. *Eur. Polym. J.* **2011**, *47*, 1081–1089. [CrossRef]
4. Wang, X.; Song, L.; Yang, H.; Lu, H.; Hu, Y. Synergistic Effect of Graphene on Antidripping and Fire Resistance of Intumescent Flame Retardant Poly(butylene succinate) Composites. *Ind. Eng. Chem. Res.* **2011**, *50*, 5376–5383. [CrossRef]
5. Wang, X.; Hu, Y.; Song, L.; Xing, W.; Lu, H.; Lv, P.; Jie, G. Flame retardancy and thermal degradation mechanism of epoxy resin composites based on a DOPO substituted organophosphorus oligomer. *Polymer* **2010**, *51*, 2435–2445. [CrossRef]
6. Slonczewski, J.C.; Weiss, P.R. Band structure of graphite. *Phys. Rev.* **1958**, *109*, 272. [CrossRef]
7. Stankovich, S.; Dikin, D.A.; Dommett, G.H.B.; Kohlhaas, K.M.; Zimney, E.J.; Stach, E.A. Graphene-based composite materials. *Nature* **2006**, *442*, 282–286. [CrossRef] [PubMed]
8. Yu, A.P.; Ramesh, P.; Itkis, M.E.; Bekyarova, E.; Haddon, R.C. Graphite nanoplatelet-epoxy composite thermal interface materials. *J. Phys. Chem. C* **2007**, *111*, 7565–7569. [CrossRef]

9. Hsiao, M.C.; Liao, S.H.; Yen, M.Y.; Liu, P.I.; Pu, N.W.; Wang, C.A.; Ma, C.C.M. Preparation of Covalently Functionalized Graphene Using Residual Oxygen-Containing Functional Groups. *ACS Appl. Mater. Interfaces* **2010**, *2*, 3092–3099. [CrossRef] [PubMed]

10. Wu, H.; Tang, B.; Wu, P. Development of novel SiO_2–GO nanohybrid/polysulfone membrane with enhanced performance. *J. Membr. Sci.* **2014**, *451*, 94–102. [CrossRef]

11. Wu, G.; Ma, L.; Liu, L.; Chen, L.; Huang, Y. Preparation of SiO_2–GO hybrid nanoparticles and the thermal properties of methylphenyl silicone resins/SiO_2–GO nanocomposites. *Thermochim. Acta* **2015**, *613*, 77–86. [CrossRef]

12. Ferreira, F.V.; Brito, F.S.; Franceschi, W.; Simonetti, E.A.N.; Cividanes, L.S.; Chipara, M.; Lozano, K. Functionalized graphene oxide as reinforcement in epoxy based Nanocomposites. *Surf. Interfaces* **2018**, *10*, 100–109. [CrossRef]

13. Ferreira, F.V.; Cividanes, L.D.S.; Brito, F.S.; de Menezes, B.R.C.; Franceschi, W.; Simonetti, E.A.N.; Thim, G.P. *Functionalization of Carbon Nanotube and Applications, Functionalizing Graphene and Carbon Nanotubes*; Springer International Publishing: New York, NY, USA, 2016; pp. 31–61.

14. Phiri, J.; Gane, P.; Maloney, T.C. General overview of graphene: Production, properties and application in polymer composites. *Mater. Sci. Eng. B* **2017**, *215*, 9–28. [CrossRef]

15. Feng, Y.; He, C.; Wen, Y.; Ye, Y.; Zhou, X.; Xie, X.; Mai, Y.W. Improving thermal and flame retardant properties of epoxy resin by functionalized graphene containing phosphorous, nitrogen and silicon elements. *Compos. Part A Appl. Sci. Manuf.* **2017**, *103*, 74–83. [CrossRef]

16. Yuan, B.; Hu, Y.; Chen, X.; Shi, Y.; Niu, Y.; Zhang, Y.; He, S.; Dai, H. Dual modification of graphene by polymeric flame retardant and $Ni(OH)_2$ nanosheets for improving flame retardancy of polypropylene. *Compos. Part A Appl. Sci. Manuf.* **2017**, *100*, 106–117. [CrossRef]

17. Luo, J.; Yang, S.; Lei, L.; Zhao, J.; Tong, Z. Toughening, synergistic fire retardation and water resistance of polydimethylsiloxane grafted graphene oxide to epoxy nanocomposites with trace phosphorus. *Compos. Part A Appl. Sci. Manuf.* **2017**, *100*, 275–284. [CrossRef]

18. Park, S.J.; Jin, F.L.; Park, J.H.; Kim, K.S. Synthesis of a novel siloxane-containing diamine for increasing flexibility of epoxy resins. *Mater. Sci. Eng. A* **2005**, *399*, 377–381. [CrossRef]

19. Liao, S.H.; Liu, P.L.; Hsiao, M.C.; Teng, C.C.; Wang, C.A.; Ger, M.D.; Chiang, C.L. One-Step Reduction and Functionalization of Graphene Oxide with Phosphorus-Based Compound to Produce Flame-Retardant Epoxy Nanocomposite. *Ind. Eng. Chem. Res.* **2012**, *51*, 4573–4581. [CrossRef]

20. Abdullah, S.I.; Ansari, M.N.M. Mechanical properties of graphene oxide(GO)/epoxy composites. *HBRC J.* **2015**, *11*, 151–156. [CrossRef]

21. Zeng, Y.; Zhou, Y.; Kong, L.; Zhou, T.; Shi, G. A novel composite of SiO_2-coated graphene oxide and molecularly imprinted polymers for electrochemical sensing dopamine. *Biosens. Bioelectron.* **2013**, *45*, 25–33. [CrossRef] [PubMed]

22. Yu, B.; Wang, X.; Qian, X.; Xing, W.; Yang, H.; Ma, L.; Lin, Y.; Jiang, S.; Song, L.; Hu, Y.; Lo, S. Functionalized graphene oxide/phosphoramide oligomer hybrids flame retardant prepared via in situ polymerization for improving the fire safety of polypropylene. *RSC Adv.* **2014**, *60*, 31782–31794. [CrossRef]

23. Zhang, L.; Li, Y.; Zhang, L.; Li, D.W.; Karpuzov, D.; Long, Y.T. Electrocatalytic Oxidation of NADH on Graphene Oxide and Reduced Graphene Oxide Modified Screen-Printed Electrode. *Int. J. Electrochem. Sci.* **2011**, *6*, 819–829.

24. Huang, G.; Chen, S.; Tang, S.; Gao, J. A novel intumescent flame retardant-functionalized graphene: Nanocomposite synthesis, characterization, and flammability properties. *Mater. Chem. Phys.* **2012**, *135*, 938–947. [CrossRef]

25. Hu, W.; Yu, B.; Jiang, S.D.; Song, L.; Hu, Y.; Wang, B. Hyper-branched polymer grafting graphene oxide as an effective flame retardant and smoke suppressant for polystyrene. *J. Hazard. Mater.* **2015**, *300*, 58–66. [CrossRef] [PubMed]

26. Gao, F.; Tong, L.; Fang, Z. Effect of a novel phosphorous nitrogen containing intumescent flame retardant on the fire retardancy and the thermal behavior of poly(butylene terephthalate). *Polym. Degrad. Stab.* **2006**, *91*, 1295–1299. [CrossRef]

27. Doyle, C.D. Estimating thermal stability of experimental polymers by empirical thermogravimetric analysis. *Anal. Chem.* **1961**, *33*, 77–79. [CrossRef]

28. Park, S.H.; Lee, S.G.; Kim, S.H. Thermal Decomposition Behavior of Carbon Nanotube Reinforced Thermotropic Liquid Crystalline Polymers. *J. Appl. Polym. Sci.* **2011**, *122*, 2060–2070. [CrossRef]
29. Park, S.J.; Cho, M.S. Thermal stability of carbon-MoSi$_2$-carbon composites by thermogravimetric analysis. *J. Mater. Sci.* **2000**, *35*, 3525–3527. [CrossRef]
30. Ozawa, T. A new Method of Analyzing Thermogravi-metric Data. *J. Therm. Anal.* **1965**, *38*, 1881–1886.
31. Abou-Okeil, A.; El-Sawy, S.M.; Abdel-Mohdy, F.A. Flame retardant cotton fabrics treated with organophosphorus polymer. *Carbohydr. Polym.* **2013**, *92*, 2293–2298. [CrossRef] [PubMed]
32. Kiliaris, P.; Papaspyrides, C.D. Polymer/layered silicate (clay) nanocomposites: An overview of flame retardancy. *Prog. Polym. Sci.* **2010**, *35*, 902–958. [CrossRef]
33. Chapkin, W.A.; McNerny, D.Q.; Aldridge, M.F.; He, Y.; Wang, W.; Kieffer, J.; Taub, A.I. Real-time assessment of carbon nanotube alignment in a polymer matrix under an applied electric field via polarized Raman spectroscopy. *Polym. Test.* **2016**, *56*, 29–35. [CrossRef]
34. Silva, K.C.; Corio, P.; Santos, J.J. Characterization of the chemical interaction between single-walled carbon nanotubes and titanium dioxide nanoparticles by thermogravimetric analyses and resonance Raman spectroscopy. *Vib. Spectrosc.* **2016**, *86*, 103–108. [CrossRef]

© 2018 by the authors. Licensee MDPI, Basel, Switzerland. This article is an open access article distributed under the terms and conditions of the Creative Commons Attribution (CC BY) license (http://creativecommons.org/licenses/by/4.0/).

Journal of
composites science

MDPI

Article

Reduced Graphene Oxide: Effect of Reduction on Electrical Conductivity

Sanjeev Rao [1,2,*], Jahnavee Upadhyay [1], Kyriaki Polychronopoulou [3], Rehan Umer [2] and Raj Das [1]

[1] Centre for Advanced Composite Materials, Department of Mechanical Engineering, The University of Auckland, Auckland Mail Centre, Auckland 1142, New Zealand; s.rao@auckland.ac.nz (J.U.); r.das@auckland.ac.nz (R.D.)

[2] Areospace Research and Innovation Centre (ARIC), Khalifa University of Science, Technology and Research, P.O. Box, Abu Dhabi 127788, UAE; rehan.umer@kustar.ac.ae

[3] Department of Mechanical Engineering, Khalifa University of Science, Technology and Research, P.O. Box, Abu Dhabi 127788, UAE; kyriaki.polychrono@kustar.ac.ae

* Correspondence: sanjeev.rao@kustar.ac.ae; Tel.: +971-2-5018509

Received: 5 March 2018; Accepted: 4 April 2018; Published: 9 April 2018

Abstract: In this study, the effect of reduction on the electrical conductivity of Graphene Oxide (GO) is investigated. The aim of this fabrication was to render electromagnetic interference (EMI) shielding to thin polymer films using GO as fillers. The electrical conductivity was determined using the four-probe method and shielding effectiveness was theoretically determined using the experimentally obtained conductivity values. The initial oxidation of graphite was performed using Hummers' method and the oxidized GO was dispersed in water for further exfoliation by ultrasonication. Thin films of sonicated GO dispersions were solution casted and dried in a convection oven at 50 °C overnight. The dried films were treated with 48% hydrobromic acid (HBr), 95% hydrochloric acid (HCl) or 66% hydroiodic acid (HI) for 2 h, 24 h or 48 h. A partial factorial design of experiments based on Taguchi method was used to identify the best reducing agent to obtain maximum electrical conductivity in the partially reduced GO films. The experimental analysis indicates that the electrical resistivity of GO is highly dependent on the type of acid treatment and the samples treated with HI acid exhibited lowest resistivity of ~0.003 $\Omega \cdot$cm. The drop in resistivity value after chemical reduction was of the order of 10,000 times, and range obtained in this work is among the lowest reported so far. The theoretical EMI shielding of the reduced GO film provided a shielding effectiveness of 5.06 dB at 12 GHz.

Keywords: graphene oxide; FTIR; EMI; Taguchi; electrical conductivity

---

1. Introduction

With recent advances in material science, nanocomposites with exfoliated nanofillers have attracted considerable attention due to the superior mechanical and functional properties offered by the fillers at relatively lower filler loadings. Exfoliated graphene oxide (GO) as filler in particular has received considerable attention because of its mechanical and functional properties that are comparable to those of single-layered graphene with the ease of synthesis compared to that of graphene [1–3].

Considering the proximity of GO to graphene, polymer-GO nanocomposites have potential applications in memory devices and energy storage cells where the low conductivity of GO can help in creating a barrier for inter-lamellar electron-hole re-combinations. Other applications can be in field effect transistors, solar cells, energy storage devices, etc. Additionally, they appear to improve the resistance to gas permeability, suggesting potential for applications in food and beverage, packaging, and filtration applications [1,4–7]. An interesting example where the conductive properties of the polymer-GO composites can be utilized is in ElectroMagnetic wave Interference (EMI) shielding and are currently emerging is a new class of materials for broad-band electromagnetic absorption [1,3,7–16].

GO is inherently non-conductive because of the absence of percolating pathways between sp^2 carbon clusters which is indeed a carrier transport mechanism in the graphene. A single sheet of graphene oxide consists of a hexagonal ring-based carbon network having both sp^2-hybridised and sp^3-hybridised carbon atoms, bearing hydroxyl and epoxide functional groups on either side of the sheet, whereas the sheet edges are mostly occupied by carboxyl and carbonyl groups [17–21], meaning that the functional groups attached to the plane influence its conductivity, while functional groups attached to the edge may not. GO as a whole on the contrary has large chemical functional groups attached to the carbon plane with several structural defects within the plane, both of which may severely decrease the electrical conductivity. However, it can be restored by varying their synthesis and processing, as they are non-stoichiometric in nature. A common approach is to attach polymers or polymerization initiators on to the GO platelets and reduce the oxidized surface [4]. This reduction process must aim at eliminating only the epoxy and hydroxyl groups on the plane, while retaining other groups, such as carboxyl, carbonyl, and ester groups at the edges.

There are different routes to synthesize GO for conductivity [22], out of which thermal and chemical routes are widely accepted to restore electrical conductivity. In the thermal reduction method, GO is exposed to temperature ramp rates greater than 2000 °C/min up to about 1050 °C, in effect splitting the graphite oxide into individual sheets through evolution of CO_2. The thermal shock decomposes the oxygen-containing groups attached to the carbon plane, and the escape of the resultant gases exfoliate the stacked graphene layers [5,9,11,23]. However, this process is suitable for bulk production, the yield is usually small, and the quality is poor with wrinkling and severe lattice defects. In the chemical route, GO is reduced using chemical reagents based on their chemical reactions with GO. Hydrazine is universally accepted as a good chemical reagent to reduce GO [2,17,24,25]. The reduction by hydrazine and its derivatives, such as hydrazine hydrate and dimethylhydrazine [12], is as simple as adding liquid reagents to aqueous dispersion of GO and drying it, wherein the reagents increase the hydrophobicity that results in agglomerated graphene-based nanosheets that may be electrically conductive. One drawback with this process is that the hydrazine agents usually tend leave C-N groups behind, in the form of hydrazones, amines, aziridines, etc., which may have detrimental effect on the electronic structure and properties of reduced GO [26,27]. However, by finely tuning the reduction parameters, it is possible to get rid of the C-N groups and indeed obtain electrically conductive GO. Therefore, in view of tuning the process parameters to achieve electrical conductivity, a systematic study on the effect of varying the process parameters on the electrical conductivity of the reduced GO discussed in this work.

For EMI shielding, conducting fillers in the form of nanocomposites are highly suitable due to ease of processing at relatively low costs with the benefit of various mechanical and functional properties of polymers. Polymer nanocomposites with carbon-based fillers have been extensively investigated in various works [23–27] with results showing their performance to be dependent on the type of matrix and filler [28,29], structural morphology and dispersion [26,27], processing method, matrix-filler interaction, etc. The EMI shielding effectiveness (EMI SE) of composite materials depends upon intrinsic conductivity of filler, its aspect ratio [30] and conductivity of the composite [31–34]. Therefore, primarily, EMI SE is a function of conductivity, but as the filler loading increases, the shielding will depend on the thickness of the specimen too, due to increased layering because of higher loading [30,35]. However, beyond percolation thresholds, the EMI shielding effectiveness also rises rapidly with the increase in electrical conductivity. However, the primary mechanism in such case is reflection due to interaction of incident radiation with the free electrons on the shield surface while absorption is the secondary mechanism. Therefore, the EMI SE can be tailored by varying the filler concentrations and their dispersion throughout the media. However, the medium of shielding effectiveness (absorption or reflection) may indeed vary accordingly.

In this study, a relatively novel chemical reduction method that uses hydrohalic acids for removal of oxygen-containing groups from GO, has been developed to tune the electrical conductivity of GO. Factorial design of experiments based on the Taguchi method was used to identify the key processing parameters and their synergistic effects, affecting the electrical property of the GO films. The post-reduction characterization of GO was performed using X-ray diffractometry, Fourier transfer infrared imaging and scanning electron microscopy.

2. Materials and Methods

Graphite flakes obtained from Sigma-Aldrich (Auckland, New Zealand) were powdered to particle size of 30 μm. 95–97% Sulfuric Acid (H_2SO_4) from Merck Group (Darmstadt, Germany) Potassium Permanganate ($KMnO_4$) from Ajax Chemicals, Australia, Sodium Nitrate $NaNO_3$, 30% Hydrogen Peroxide (H_2O_2), 95% Hydrochloric acid (HCl), 48% Hydrobromic Acid (HBr) and 66% Hydroiodic acid (HI) from Ajax Finechem Inc. (Auckland, New Zealand) were used as-received. *N,N*-dimethylformamide (DMF, Anhydrous, 99.8%) and tetrahydrofuran (THF) were bought from Sigma-Aldrich and ECP Ltd. (Romil, Cambridge, UK), respectively, and the polymer matrix PMMA was bought from Chi Mei Corporation (Tainan, Taiwan).

Wide angle X-ray scattering (WAXS) technique was used to evaluate the crystallinity and inter-layer distance between adjacent layers of the multi-layered material. Fourier transfer infrared spectroscopy, FT-IR (Nicolet 8700, ATR-FTIR, Waltham, Massachusetts, USA), was used to confirm the presence/absence of oxygen-containing groups in GO after oxidization of graphite and partial reduction of GO, respectively. Scanning electron microscope, SEM (Philips XL30S FEG with Gatan Alto Cryo-trans System, Auckland, New Zealand), was used to visualize the GO sheets and nanofibers after platinum sputtering. The four-probe method developed in-house was used to measure the electrical conductivity of the GO sheets.

2.1. Graphene Oxide

Graphene oxide was synthesized using modified Hummers' method [29] and the details of the method can be found in the reference as listed, and pictures at various stages of chemical processing is shown in Figure 1. To achieve exfoliation, the GO obtained from the above process was dispersed in water and subjected ultrasonication. Thin films of sonicated GO dispersions were prepared by casting the mixture on petri dishes followed by drying in the oven at 50 °C, overnight.

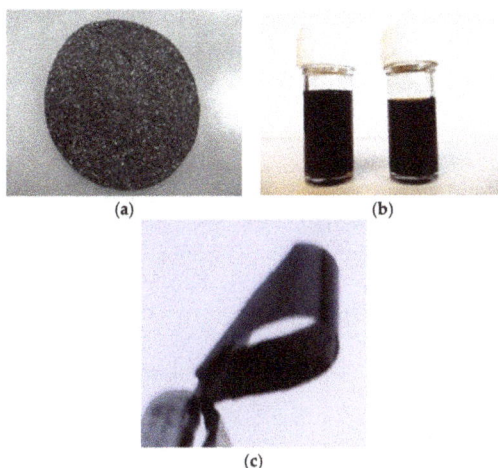

(a) (b)

(c)

Figure 1. Stages of chemical processing of GO (a) Pulverized graphite that was used for reduction (b) GO dispersed in water after synthesis using Hummers' method (c) GO thin film.

2.2. X-Ray Diffractometry (XRD)

The X-ray diffraction spectra of graphite powder and GO samples exhibit a sharp peak at $2\theta = 11.2°$ compared to the strong characteristic peak at $2\theta = 26.4°$ for graphite (Figure 2). By using Debey-Sherrer's equation [30], the inter-layer distance for GO sheets was calculated to be 0.82 nm, which is greater than that (0.34 nm) for graphene layers. This increase in the interlayer distance is attributed to the presence of water (H_2O) molecules and various oxygen-containing functional groups as identified by FTIR. This agrees with [36]. These species do facilitate the hydration and exfoliation of the GO sheets. Therefore, the plots in Figure 2 indicate the effectiveness of the modified Hummers' method in oxidizing the graphite leading to increased inter-layer distance between the graphene layers.

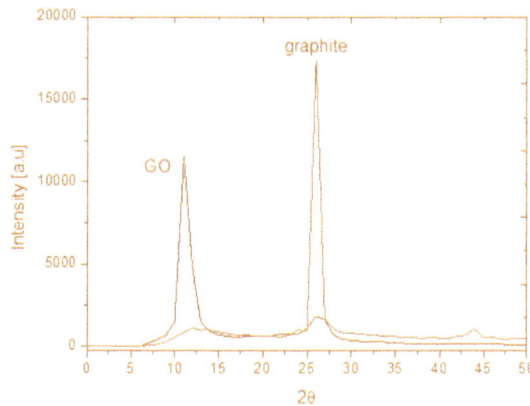

Figure 2. XRD traces for Graphite and GO confirming the synthesis.

It can be observed that the peak at $2\theta = 11.2°$ exhibits higher intensity for the GO material, indicating a higher amount of water molecules/oxygen functional groups present in the basal plane, thus improving the extent of oxidation and reflecting the influence of increased H_2SO_4 in the acidic media. In the case of graphite, this peak is appeared with minor intensity, close to the background of the pattern. The small peaks at $2\theta = 26.4$ degrees for GO sample indicates the presence of unconverted graphite and can be related to the efficiency of the process. The presence of unconverted graphite and the difference in the extent of oxidation were also discerned in the thermal tests in the form of exothermic peaks and varying heat flux for oxidation reactions.

The GO sheets obtained after drying usually exhibit high electrical resistivity but by treating them with hydrohalic acids reducing agents, recovery to an extent is possible. The extent of exfoliation during the reduction of graphite to GO, and the duration of chemical treatment, play a vital role in determining the extent of electrical conductivity restoration. Therefore, an optimum combination of those parameters is paramount to the electrical property restoration.

Investigation of such a complex, multivariable systems and the influence of various parameters on the electrical properties of the reduced GO films individually tend to become tedious, and hence it is difficult to assess the synergistic effects between the parameters. Therefore, partial factorial design of experiments (DoE), accompanied by statistical analysis based on Taguchi method, was used to obtain the best possible combination of these parameters in a minimum number experimental trials.

2.3. Design of Experiments

Three levels of ultrasonication time, reducing agent type and reaction duration were selected (Table 1) to construct an L_9 orthogonal array. Their levels and different combinations are listed in Table 2.

Table 1. Factors for Taguchi L$_9$ matrix.

Factor	Level 1	Level 2	Level 3
Sonication Time (A)	1 h	10 h	20 h
Reducing Agent (B)	Hydrochloric acid	Hydrobromic Acid	Hydroiodic Acid
Reaction Time (C)	2 h	24 h	48 h

Table 2. Experimental values of electrical resistivity for GO samples.

Standard Order No.	A	B	C	Resistivity (Ω·cm)
1	1	1	1	0.330
2	1	2	2	0.560
3	1	3	3	0.006
4	2	1	2	0.641
5	2	2	3	0.076
6	2	3	1	0.003
7	3	1	3	0.365
8	3	2	1	0.347
9	3	3	2	0.022

The graphical representation of the effect of individual factors and combination of factors towards the overall electrical resistivity is shown in Figure 3. The contribution of each factor towards the overall resistivity is estimated as the average of all the resistivity values when the factor is set at its particular level, thus minimizing any random errors.

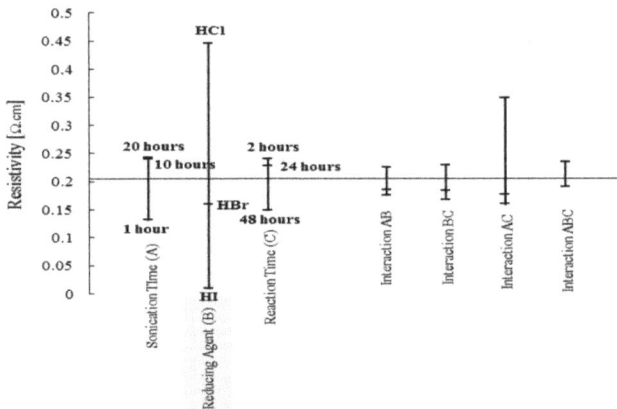

Figure 3. Factor Effect Plot for Taguchi Analysis.

In Figure 3, the overall average of resistivity from all the trials is represented as a horizontal line with the averages of all the individual factors set at their particular levels plotted on either side of it. The physical size of the line determines the factor's contribution towards the overall response of the system's electrical resistivity.

In the factor effects plot the reducing agent type plays a detrimental role in the electrical resistivity (hence increasing the conductivity) of GO. The average resistivity for GO samples treated with hydroiodic acid appear to be in the range of 0.003–0.022 Ω·cm, followed by those treated with hydrobromic acid (0.056–0.347 Ω·cm) and finally with hydrochloric acid (0.33–0.641 Ω·cm).

An increase in ultrasonication time of the GO suspension over 1 h resulted in an increased resistivity, i.e., a decrease in conductivity of the GO. This negative effect for the ultrasonication time can be attributed to the breaking up of GO sheets into smaller fragments/particulates, rather than an increase in inter-layer distance with increased amount of sonication. The SEM images of GO solutions, subjected to ultrasonication for 1 h, 10 h and 20 h, are presented in Figure 4, respectively, which shows a higher degree of fragmentation of GO sheets with increased ultrasonication.

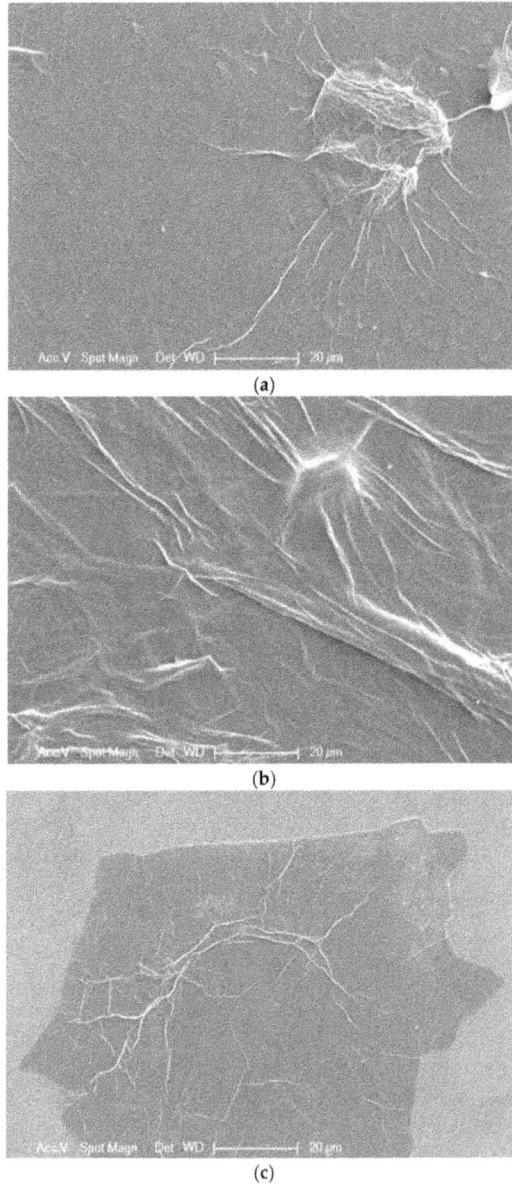

(a)

(b)

(c)

Figure 4. SEM images of GO after ultrasonication for (**a**) 1 h that shows relatively less wrinkled sheets (**b**) 10 h, showing wrinkled features; and (**c**) 20 h where it is fragmented and wrinkly.

The third factor considered was the reaction time. When the reaction time was increased from 2 h to 48 h, an increase in conductivity was observed. Therefore, an increase in reaction time enhances the overall conductivity of the film. However, the effect is relatively small compared to that of the reducing agent type, because hydroiodic acid produced resistivity as low as 0.003 Ω.cm even at lower reaction times (e.g., at 2 h). Such inferences can be made by evaluating the relevant factor interaction effects, estimated by examining the variation of each factor with respect to their levels.

To understand the synergistic or antagonistic effect of the factors on the overall response of the system, factor interaction plots were used, shown in Figure 5. The size of the lines that are deviating from the average, depict the two factor interaction *sonication time and reaction time* (AC) to contribute more towards the overall conductivity compared to the other factor interactions (*sonication time and reducing agent type* (AB), *reducing agent type and reaction time* (BC) and *sonication time and reducing agent type and reaction time* (ABC)).

In Figure 5a, the interaction graph for AC displays a drift in the average response values as the ultrasonication time increases from 1 h to 20 h. For the reaction time at its lowest level, i.e., 2 h, the resistivity value drops from 0.33 Ω·cm to 0.003 Ω·cm when the sonication time is changed from 1 h to 10 h, and then it increases to 0.347 Ω·cm with an increase in the sonication time to 20 h. For the reducing time of 24 h, the change in the sonication time from level 1 to 3 (from 1 h to 20 h) results in an antagonistic effect (an increase in resistivity from 0.056 Ω·cm to 0.641 Ω·cm), when the sonication time changes from level 1 to 2, a sudden decrease to 0.022 Ω·cm is observed. For reducing duration of 48 h, the response behavior is similar to that of 2 h with the resistivity value showing an increasing trend, from 0.006 Ω·cm to 0.076 Ω·cm and then to 0.365 Ω·cm, with successive raise in the sonication level.

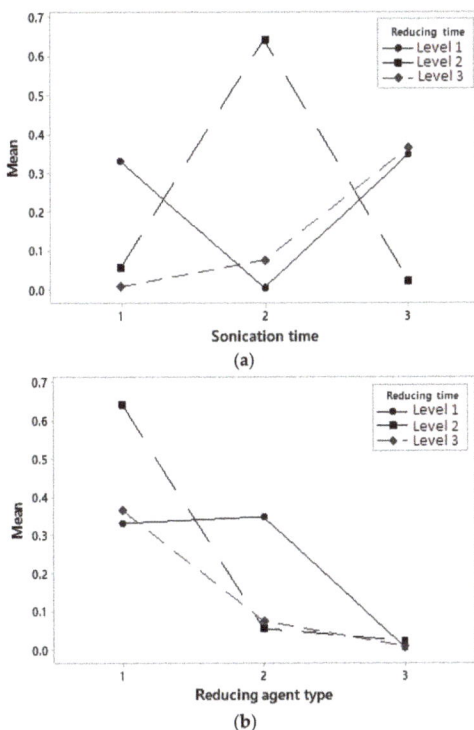

(a)

(b)

Figure 5. *Cont.*

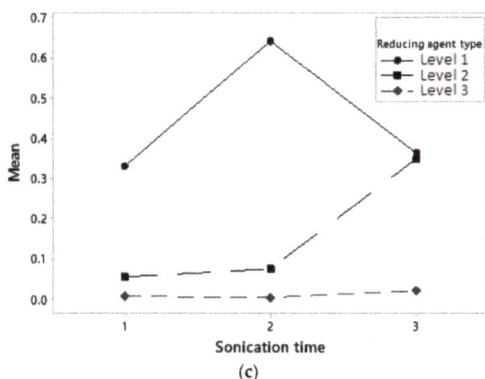

Figure 5. Synergistic and antagonistic interaction effect plots of factors (**a**) Sonication Time (A) and Reaction Time (C) (**b**) Reducing Agent (B) and Reaction Time (C) and (**c**) Sonication Time (A) and Reducing Agent (B).

The interaction graph in Figure 5b shows a change in resistivity with increase in reaction time for each of the reducing agents used (HCl, HBr and HI), with drastic decline in resistivity values observed with HI (from 0.641 $\Omega \cdot$cm to 0.056 $\Omega \cdot$cm). This indicates that the degree of exfoliation is directly proportional to the duration of sonication. For reaction time at its lowest level (2 h), the average resistivity value increases from 0.33 $\Omega \cdot$cm to 0.641 $\Omega \cdot$cm when the reducing agent is changed from HCl to HI. Therefore, for shorter sonication duration, HCL and HBr provide similar resistivity values and may be preferred over HI. At 48 h sonication, all the reducing agents provide similar exfoliation and hence resistivity values. In Figure 5c, it can be deduced from the antagonistic nature of the interaction effect, the effect of change in the sonication time is most prominent at level 1, but with escalation in the reactivity of the acid used, i.e., at levels 2 and 3 (when using HBr and HI), the effect of sonication becomes less pronounced.

3. Characterization

3.1. Fourier Transfer Infrared Spectroscopy (FTIR)

The spectra obtained by FTIR of graphite and GO samples are shown in Figure 6. A wide band at 3000–3600 cm^{-1} represents the stretching vibration due to the presence of hydroxyl group (O-H vibration). In the spectrum, the typical peaks indicating the oxygen-containing groups are visible as 1715.5 cm^{-1} for C=O (carbonyl/carboxyl), 1337.7 cm^{-1} for carboxylic C-O, 1222.9 cm^{-1} for epoxy C-O and 1060 cm^{-1} for alkoxy C-O. The peak at 1621.5 cm^{-1} corresponding to the in-plane vibrations of aromatic C=C and the skeletal vibrations of the graphene sheets corroborating for the success of oxidation of graphitic domains. The FTIR spectra of GO after treatment with hydrobromic acid (RGO in Figure 6) shows significant decrease of the bands that correspond to C-OH hydroxyl and C=O groups, while the peaks pertaining to epoxy and alkoxy C-O groups have completely disappeared. These FTIR spectra confirm that the GO samples have been significantly de-oxidized and support the hypothesis proposed by Lin et al. [22] that low-temperature reduction processes result in decomposition of carboxylic and carbonyl groups leading to less functional groups in the RGO sample. Though, some residual oxygen-containing species remain in the material.

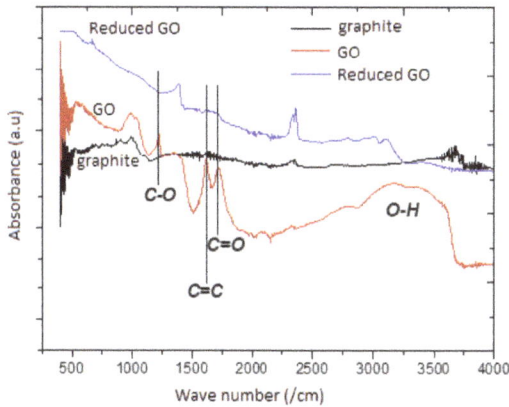

Figure 6. FTIR spectra for Graphite, GO and Reduced GO confirming the reduction process using hydrobromic acid.

3.2. Electromagnetic Shielding

Electromagnetic interference shielding is the attenuation of the incident electromagnetic waves in the material and is expressed by the mathematical formula given by [5,13,31–33].

$$SE = 10 \log \frac{P_i}{P_t} \tag{1}$$

where SE is the shielding effectiveness in decibel (dB), P_t is the power of the transmitted wave and P_i is the power of incident wave. When an electromagnetic wave is incident on a surface, it is absorbed (A), reflected (R) or transmitted (T) and the summation, A + R + T = 1 [13] and the attenuation mainly consists of the three components: Absorption A, Reflection R and Multiple Reflections M [32,33]. Hence, the total EMI SE can be expressed as summation of these three components, i.e., SE = SE_a + SE_r + SE_m.

The effect of multiple reflections is often negligible when SE_a >= 10 dB [14,32,34,35] as the intensity of an incident wave inside the material after reflection will depend on (1 − R). Therefore, Effective Absorption inside a shielding material is defined using Equation (2) and the SE component rendered by absorption and reflection phenomena can be determined by using Equations (3) and (4) respectively.

$$A_{eff} = \frac{1 - R - T}{1 - R} \tag{2}$$

$$SE_a = 10 \log \left(1 - A_{eff}\right) = 10 \log \left(\frac{T}{1 - R}\right) \tag{3}$$

$$SE_r = 10 \log(1 - R) \tag{4}$$

In dynamic fields, the eddy current behaves complexly and it is common practice to develop approximate solutions while modelling shielding enclosures. For thin films, the shielding is dominated by the magnetic field and is governed by the magnetic permeability of the shield. As frequency increases, the shieling mechanism becomes greater due to induced current density that reduces inward from the core, exponentially. The skin depth δ over which the amplitude of incident wave will be effectively reduced is defined by Equation (5) and the SE due to absorption component can be calculated by Equation (6). Therefore, the EMI shielding effectiveness of a composite specimen can be studied as a function of its electrical conductivity which in turn depends on the intrinsic conductivity of the filler, filler morphology and dispersion and the interfacial interaction between the filler and the matrix molecules [5,12,15,36].

$$\delta = \frac{1}{\sqrt{\pi f \mu \sigma}} \tag{5}$$

where σ is the electrical conductivity in S/cm, μ is the magnetic permeability of the material and is equal to 1 for non-magnetic materials and f is the frequency of incident radiation in MHz.

$$SE_a = K\frac{t}{\delta} = Kt\sqrt{\pi f \mu \sigma} \tag{6}$$

where δ is the skin depth for effective shielding in m, K is the constant for the shield material and t is the thickness of the shield in m.

The thickness of any planar shield can be represented in terms of skin depths to indicate the shielding potential of the material. In addition to the absorption over skin depth, significant reduction of the impinging electromagnetic field can occur through reflection from the shield surface. The formula of far-field reflection loss for a plane-wave radiation is given by Equation (7) [19].

$$SE_r = 108 + \log\left(\frac{\sigma}{f\mu}\right) \tag{7}$$

The maximum value of conductivity attained through chemical reduction of GO sheets was 39,000 S/m. The thickness of the GO films was 0.1 mm and the predicted values of EMI shielding effectiveness for GO thin films are listed in Table 3.

Table 3. EMI SE calculation for reduced GO specimens using hydrobromic acid.

Sr. No.	Electrical Conductivity (S/m)	Frequency (GHz)	EMI Shielding Effectiveness, dB (SE_a)
1	39,000	8	4.13
2	39,000	9	4.38
3	39,000	10	4.62
4	39,000	11	4.84
5	39,000	12	5.06

4. Conclusions

In this paper, GO synthesized from graphite using the modified Hummers' and Offeman methods was characterized using X-ray diffractometry, Fourier transfer infrared imaging and scanning electron microscopy. The XRD results affirmed the increase in inter-layer distance between graphene layers, and FTIR confirmed the presence of oxygen-containing reactive groups in the oxidized graphite samples. The SEM images revealed typical wrinkled and wavy morphology of GO thin sheets at high magnifications.

A simple, hazard-free, and effective low-temperature chemical reduction method was developed with relatively novel use of hydrohalic acids for removal of oxygen-containing groups from GO. A partial factorial DoE along with Taguchi analysis was used to obtain the best possible combination of parameters, i.e., ultrasonication time, reducing agent and reaction time, to improve the electrical conductivity of reduced GO films. The Taguchi analysis of factorial effects revealed that the effectiveness of the reduction process is primarily dependent on the reducing agent used. Immersion of GO thin films in strong hydrohalic acids, such as hydroiodic acid, can effectively reduce GO even when the reaction time and level of sonication are very low. The post-reduction electrical resistivity values for GO thin films, as measured using the four-point probe method, demonstrated a high degree of improvement in conductivity. The drop in resistivity value after chemical reduction was of the order of 10,000 times, and the measured values were in the range of 0.003–0.641 Ω.cm. The electrical resistivity range obtained in this work is among the lowest reported so far. This suggests the vast

J. Compos. Sci. **2018**, *2*, 25

potential of GO-based nanocomposites in applications pertaining to electronic field, where GO could be tuned to exhibit insulating, semiconducting, or conducting behavior, as per the requirement.

Acknowledgments: The authors thank Prateek Jaiswal for assistance with the experimental work and Dong Liu for her assistance with chemical synthesis.

Author Contributions: This work is a part of Jahnavee Upadhyay's ME research at the University of Auckland, and she along with Sanjeev Rao and Raj Das conceived and designed the experiments; Jahnavee Upadhyay under Sanjeev Rao's supervision performed the experiments; Rehan Umer and Kyriaki Polychronopoulou have contributed in analyzing the data; The University of Auckland and Khalifa University of Science and Technology contributed reagents/materials/analysis tools for this work; Sanjeev Rao wrote the paper.

Conflicts of Interest: The authors declare no conflict of interest.

References

1. Kuilla, T.; Bhadra, S.; Yao, D.; Kim, N.H.; Bose, S.; Lee, J.H. Recent Advances in graphene based polymer composites. *Prog. Polym. Sci.* **2010**, *35*, 1350–1375. [CrossRef]
2. Liang, J.; Wang, Y.; Huang, Y.; Ma, Y.; Liu, Z.; Cai, J.; Zhang, C.; Gao, H.; Chen, Y. Electromagnetic interference shielding of graphene/epoxy composites. *Carbon* **2009**, *47*, 922–925. [CrossRef]
3. Eda, G.; Unalan, H.E.; Rupensinghe, N.; Amaratunga, G.A.J.; Manish, C. Field Emission from graphene based composite thin films. *Appl. Phys. Lett.* **2008**, *93*, 233502. [CrossRef]
4. Senguptaa, R.; Bhattacharyaa, M.; Bandyopadhyay, S.; Bhomicka, A.K. *A Review on the Mechanical and Electrical Properties of Graphite and Modified Graphite Reinforced Polymer Composites*; Elsevier: New York, NY, USA, 2010.
5. Kim, H.; Miura, Y.; Macosko, C.W. Graphene/Polyurethane Nanocomposites for Improved Gas Barrier and Electrical Conductivity. *Chem. Mater.* **2010**, *22*, 3441–3450. [CrossRef]
6. Compton, O.C.; Kim, S.; Pierre, C.; Torkelson, J.M.; Nguyen, S.T. Crumpled Graphene Nanosheets as Highly Effective Barrier Property Enhancers. *Adv. Mater.* **2010**, *22*, 4759–4763. [CrossRef] [PubMed]
7. Choi, W.; Lahiri, I.; Seelaboyina, R.; Kang, Y.S. Synthesis of Graphene and Its Applications: A Review. *Solid State Mater. Sci.* **2010**, *35*, 52–71. [CrossRef]
8. Yan, J.; Wei, T.; Shao, B.; Fan, Z.; Qian, W.; Zhang, M.; Wei, F. Preparation of a graphene nanosheet/polyaniline composite with high specific capacitance. *Carbon* **2010**, *48*, 487–493. [CrossRef]
9. Potts, J.R.; Dreyer, D.R.; Bielawski, C.W.; Ruoff, R.S. *Graphene-Based Polymer Nanocomposites*; Elsevier: New York, NY, USA, 2010.
10. Bao, Q.; Zhang, H.; Yang, J.-X.; Wang, S.; Tang, D.Y.; Jose, R.; Ramakrishna, S. Lim, C.T.; Loh, K.P. Graphene-Polymer Nanofiber membrane for ultrafast photonics. *Adv. Funct. Mater.* **2010**, *20*, 782–791. [CrossRef]
11. Ansari, S.; Giannelis, E.P. Functionalized Graphene Sheet—Poly(vinylidene fluoride) Conductive Nanocomposites. *J. Polym. Sci.* **2009**, *47*, 888–897. [CrossRef]
12. Stankovich, S.; Dikin, D.A.; Dommett, G.H.B.; Kolhaas, K.M.; Zimney, E.J.; Stach, E.A.; Piner, R.D.; Nguyen, S.T.; Ruoff, R.S. Graphene-based composite materials. *Nature* **2006**, *442*, 282–286. [CrossRef] [PubMed]
13. Vovchenko, L.L.; Matzui, L.Y.; Oliynyk, V.V.; Launetz, V.L. The Effect of Filler Morphology and Distribution on Electrical and Shielding Properties of Graphite-Epoxy Composites. *Mol. Cryst. Liq. Cryst.* **2011**, *535*, 179–188. [CrossRef]
14. Cao, M.-S.; Song, W.-L.; Hou, Z.-L.; Wen, B.; Yuan, J. The effects of temperature and frequency on the dielectric properties, electromagnetic interference shielding and microwave-absorption of short carbon fiber/silica composites. *Carbon* **2010**, *48*, 788–796. [CrossRef]
15. Yang, Y.L.; Gupta, M.C. Novel carbon nanotube-polystyrene foam composites for electromagnetic interference shielding. *Nano Lett.* **2005**, *5*, 2131–2134. [CrossRef] [PubMed]
16. Che, R.C.; Peng, L.M.; Duan, X.F.; Chen, Q.; Liang, X.L. Microwave absorption enhancement and complex permittivity and permeability of Fe encapsulated within carbon nanotubes. *Adv. Mater.* **2004**, *16*, 401–405. [CrossRef]
17. Park, S.; Ruoff, R.S. Chemical methods for the production of graphenes. *Nat. Nanotechnol.* **2009**, *4*, 217–224. [CrossRef] [PubMed]

18. Li, D.; Mueller, M.B.; Gilje, S.; Kaner, R.B.; Wallace, G.G. Processable aqueous dispersions of graphene nanosheets. *Nat. Nanotechnol.* **2008**, *3*, 101–105. [CrossRef] [PubMed]

19. Li, D.; Kaner, R.B. Materials science—Graphene-based materials. *Science* **2008**, *320*, 1170–1171. [CrossRef] [PubMed]

20. Lerf, A.; He, H.Y.; Forster, M.; Klinowski, J. Structure of graphite oxide revisited. *J. Phys. Chem. B* **1998**, *102*, 4477–4482. [CrossRef]

21. Yang, Y.; Wang, J.; Zhang, J.; Liu, J.; Yang, X.; Zhao, H. Exfoliated Graphite Oxide Decorated by PDMAEMA Chains and Polymer Particles. *Langmuir* **2009**, *25*, 11808–11814. [CrossRef] [PubMed]

22. Lin, Z.; Yao, Y.; Li, Z.; Liu, Y.; Li, Z.; Wong, C.-P. Solvent-Assisted Thermal Reduction of Graphite Oxide. *J. Phys. Chem. C* **2010**, *114*, 14819–14825. [CrossRef]

23. Saito, H.; Inoue, T. Light and X-Ray Scatterings. In *Polymer Characterisation Techniques and Their Application to Blends*; Simon, G.P., Ed.; Oxford University Press: Washington, DC, USA, 2004; pp. 313–345.

24. Dikin, D.A.; Stankovich, S.; Zimney, E.J.; Piner, R.D.; Dommett, G.H.B.; Evmenenko, G.; Nguyen, S.T.; Ruoff, R.S. Preparation and characterization of graphene oxide paper. *Nature* **2007**, *448*, 457–460. [CrossRef] [PubMed]

25. Kang, H.; Kulkarni, A.; Stankovich, S.; Ruoff, R.S.; Baik, S. Restoring electrical conductivity of dielectrophoretically assembled graphite oxide sheets by thermal and chemical reduction techniques. *Carbon* **2009**, *47*, 1520–1525. [CrossRef]

26. Dreyer, D.R.; Park, S.; Bielawski, C.W.; Ruoff, R.S. The chemistry of Graphene Oxide. *Chem. Soc. Rev.* **2010**, *39*, 228–240. [CrossRef] [PubMed]

27. Jang, J.Y.; Kim, M.S.; Jeong, H.M.; Shin, C.M. Graphite oxide/poly(methyl methacrylate) nanocomposites prepared by a novel method utilizing macroazoinitiator. *Compos. Sci. Technol.* **2009**, *69*, 186–191. [CrossRef]

28. Kilic, A.; Oruc, F.; Demir, A. Effects of polarity on electrospinning process. *Text. Res. J.* **2008**, *78*, 532–539. [CrossRef]

29. Hummers, W.S.; Offeman, R.E. Preperation of Graphitic Oxide. *J. Am. Chem. Soc.* **1958**, *80*, 1339. [CrossRef]

30. Cullity, B.D. *Elements of X-Ray Diffraction*; Addison-Wesley Pub. Co.: Reading, MA, USA, 1956.

31. Tesche, F.M.; Ianoz, M.V.; Karlsson, T. *EMC Analysis Methods and Computational Models*; John-Wiley & Sons. Inc.: New Jersey, NJ, USA, 1997.

32. Wang, W.P.; Pan, C.Y. Preparation and characterization of poly(methyl methacrylate)-intercalated graphite oxide/poly(methyl methacrylate) nanocomposite. *Polym. Eng. Sci.* **2004**, *44*, 2335–2339.

33. Pande, S.; Singh, B.P.; Mathur, R.B.; Dhami, T.L.; Saini, P.; Dhawan, S.K. Improved Electromagnetic Interference Shielding Properties of MWCNT-PMMA Composites Using Layered Structures. *Nanoscale Res. Lett.* **2009**, *4*, 327–334. [CrossRef] [PubMed]

34. Al-Saleh, M.H.; Sundararaj, U. Electromagnetic interference shielding mechanisms of CNT/polymer composites. *Carbon* **2009**, *47*, 1738–1746. [CrossRef]

35. Park, K.Y.; Lee, S.E.; Kim, C.G.; Han, J.H. Fabrication and electromagnetic characteristics of electromagnetic wave absorbing sandwich structures. *Compos. Sci. Technol.* **2006**, *66*, 576–584. [CrossRef]

36. Kim, H.M.; Kim, K.; Lee, C.Y.; Joo, J.; Cho, S.J.; Yoon, H.S.; Pejaković, D.A.; Yoo, J.W.; Epstein, A.J. Electrical conductivity and electromagnetic interference shielding of multiwalled carbon nanotube composites containing Fe catalyst. *Appl. Phys. Lett.* **2004**, *84*, 589–591. [CrossRef]

© 2018 by the authors. Licensee MDPI, Basel, Switzerland. This article is an open access article distributed under the terms and conditions of the Creative Commons Attribution (CC BY) license (http://creativecommons.org/licenses/by/4.0/).

Journal of
composites science

MDPI

Article

Development of Pb-Free Nanocomposite Solder Alloys

Animesh K. Basak [1,*], Alokesh Pramanik [2], Hamidreza Riazi [3], Mahyar Silakhori [4] and Angus K. O. Netting [1]

[1]   Adelaide Microscopy, the University of Adelaide, Adelaide, SA 5005, Australia;
      angus.netting@adelaide.edu.au
[2]   Department of Mechanical Engineering, Curtin University, Bentley, WA 6845, Australia;
      alokesh.pramanik@curtin.edu.au
[3]   Department of Materials Engineering, Isfahan University of Technology, Isfahan 83714, Iran;
      h.riazi@ma.iut.ac.ir
[4]   School of Mechanical Engineering, the University of Adelaide, Adelaide, SA 5005, Australia;
      mahyar.silakhori@adelaide.edu.au
*   Correspondence: Animesh.basak@adelaide.edu.au

Received: 23 March 2018; Accepted: 16 April 2018; Published: 20 April 2018

Abstract: As an alternative to conventional Pb-containing solder material, Sn–Ag–Cu (SAC) based alloys are at the forefront despite limitations associated with relatively poor strength and coarsening of grains/intermetallic compounds (IMCs) during aging/reflow. Accordingly, this study examines the improvement of properties of SAC alloys by incorporating nanoparticles in it. Two different types of nanoparticles were added in monolithic SAC alloy: (1) Al_2O_3 or (2) Fe and their effect on microstructure and thermal properties were investigated. Addition of Fe nanoparticles leads to the formation of $FeSn_2$ IMCs alongside Ag_3Sn and Cu_6Sn_5 from monolithic SAC alloy. Addition of Al_2O_3 nano-particles do not contribute to phase formation, however, remains dispersed along primary β-Sn grain boundaries and act as a grain refiner. As the addition of either Fe or Al_2O_3 nano-particles do not make any significant effect on thermal behavior, these reinforced nanocomposites are foreseen to provide better mechanical characteristics with respect to conventional monolithic SAC solder alloys.

Keywords: Pb free; solder alloy; nanocomposite; microstructure

1. Introduction

The expansion of electronic packaging system in the context of miniaturization requires a number of factors to be taken into consideration such as wafer design, resist system, metal deposition techniques etc. The most crucial among these factors is 'soldering', without which none of the electronic packaging is complete [1,2]. Soldering as a word doesn't confine itself only in electronic industries, but also frequently used in plumbing (acid core) and sheet metal joining. Usually, the alloys that melt (using fluxing materials such as CaF_2) within the 90–450 °C are termed solder materials [2]. However, for electronic packaging, the melting temperature of solder alloy remains within 170–190 °C [2]. Lead (Pb) based alloys have a long history as solder alloys and over past decades have been the most predominant solder alloy used. However, the environmental concern strictly prohibits the use of such toxic alloys paving the way for the development of Pb-free solder alloys. Usually, whenever a set of metals are mixed to form solder alloys, a number of facts come into consideration including cost compatibility, ease of fabrication, compliance with physical, chemical, mechanical and electrical properties with substrate and finally long term environmental sustainability. It is understood that solder materials in electronic packaging undergo thermal stress, not only during fabrication and reflowing, but also at service. Therefore, it is highly probable that this thermal stress renders the solder materials unsuitable under adverse condition where the possibility of metallurgical change is eminent. Thus, the applicability of a solder alloy is attributed to its resistance under low cycle thermal fatigue which is governed by its underlying microstructure.

To fulfil such requirements, Sn–Ag–Cu (Tin–Silver–Copper) based alloys provide a good alternative [1,2]; as this ternary alloy system yields comparable mechanical, thermal, creep and fatigue properties as that of conventional Pb-based solder alloys, thanks to its eutectic microstructure. The foreseen success of these alloys is due to intermetallic compounds (IMCs) formation such as $Cu_{6.26}Sn_5$ (hexagonal), Cu_6Sn_5 (monoclinic), Ag_4Sn (hexagonal) and Ag_3Sn (orthorhombic) in β-Sn (tetragonal) matrix that provide the mechanical strength while retaining comparatively low melting point and formability [3–5]. Based on nanoindentation it was reported that failure mode of Cu_6Sn_5 and Cu_3Sn IMCs are brittle in nature in contrast of ductile mode failure of Ag_3Sn [6]. The ternary SAC system was first reported in [7] and have quasi-eutectic composition of Sn-4Ag-0.5Cu. However, a number of compositional variation of these alloys has been reported [1,8,9]. Irrespective of near eutectic composition, these alloys contain IMCs dispersed in β-Sn matrix. As reported by Lehman et al. [10], mechanical durability of such alloy at service depends on relative orientation and microstructure formation of such IMCs in β-Sn matrix. It was observed that, single grained Sn joint in solder fails to balance repeated thermal stress, however, near eutectic SAC at 80 °C underwent cyclic growth twinning all through solidification process [9]. Reflowing of such SAC based solder on Cu substrate cause coarsening effect of IMCs and degrade mechanical properties. IMCs continues to grow at reflowing temperature within short period of time irrespective of SAC alloy composition. Discontinuity of such IMCs has many disadvantages such as lack of mechanical strength, failure in low magnitude thermal cycles etc. Therefore, with a view to improve mechanical property and service reliability of these solder alloys, incorporation of reinforcing elements is beneficial. Recent investigations on nanoparticles reinforced Al-matrix composite exhibits superior mechanical properties of materials [10] in terms of limiting grain growth, strength and avoid grain coarsening. Among these reinforcing elements NiO [11], Al_2O_3 [12], TiO_2 [13], pure elements such as Ni, Fe, Bi [14,15] and even rare earth elements has been reported [16] to extent its mechanical properties further without compromising thermal behavior. Regardless of type reinforcing particles, overall objectives are matrix grain refinement, favors IMCs formation, retain alloy strength during service and restrict grain/IMCs growth during fabrication/reflowing. Though there are some reports published in literature on that, however, it's still in very early stage of development and more fundamental work in that area is foreseen.

Thus, objective of the present work is to investigate the effect of nano-size Fe or Al_2O_3 addition in monolithic Sn-4.4Ag-2.6Cu alloy on its microstructure, phase evolution and thermal behavior. Towards that, an experimental based approach was adopted to make fundamental understating and further development of durable Pb-free solder nanocomposites.

2. Experimental

2.1. Materials

The common manufacturing processes for the fabrication of composite solders are usually casting or conventional powder metallurgy routes [12,17] followed by sintering. However, powder metallurgy routes following by sintering suffer from grain/IMCs growth and contain defects such as inhomogeneous microstructure due to agglomeration, segregation, gas trapping etc. To overcome that, casting method was employed in present case. A Pb-free Sn-4.4Ag-2.6Cu (wt %) solder alloy, hereby termed as SAC, was prepared by melting commercially pure (99.99%) tin, silver and copper metals in an induction furnace at about 1000 °C for 40 min under vacuum according to their nominal composition. Then the melt was cast in ingot form and re-melted again at 1000 °C for 20 min under vacuum for proper homogenization of the alloy. After that, either Al_2O_3 or Fe nanoparticles were added in monolithic SAC alloy to form nanocomposite solder alloys and subsequently, cast in to disk shaped ingots (Ø 30 mm, 2 mm thickness) and allowed to cool slowly to form ternary eutectic microstructure. Hereby, nanocomposite solder alloys reinforced with Fe is termed as SAC1 and nanocomposite solder alloy reinforced with Al_2O_3 is termed as SAC2. Al_2O_3 nanoparticles with an average particulate size of 50 nm and Fe with an average particulate size of 30 nm (as specified by the supplier, in both case) were obtained from Advanced Pinnacle Technologies, Singapore. The alumina content was 0.5 wt %, whereas Fe content was 2 wt %. The choice of Al_2O_3 nanoparticles was based on their relatively low density and high hardness, where Fe was added to favor IMCs formation. The samples were then cold mounted in epoxy resin and metallographic polishing was carried out in Struers Tegrapol (Struers, Sweden) automatic polisher with different size diamond slurry followed by final polishing in colloidal silica suspension.

2.2. Methodology

Microstructural characterization of solder alloys was carried out in FIB-SEM (Helios Nova lab 600, FEI) equipped with energy dispersive X-ray (EDX) system. Top-view SEM micrographs were taken after standard metallographic polishing as mentioned in Section 2.1 and FIB-SEM was used to mill out material to reveal cross-sectional view. To analyse the phases present in the alloy, X-ray diffraction (XRD) was carried using a monochromatic CuKα radiation (New D8 advance, Bruker, Germany). Thermal analysis of the alloys was carried out with thermogravimetric analysis and differential scanning calorimeter (TGA/DSC 2, Mettler Toledo, Columbus, OH, USA) in an inert atmosphere (Ar) with heating rate of 5 °C/min from 50 °C to 400 °C. 10 mg of the crushed (powdered) alloy was placed in alumina crucible inside the chamber. The system was computer controlled and provided real time data.

3. Result and Discussion

3.1. Microstructure of Solder Alloys

SEM micrographs, both top-view and cross-section view, of as-cast SAC alloy is shown in Figure 1 including elemental analysis (EDX spectra). The microstructure is composed of β-Sn grains and eutectic regions that became distinguishable as a result of 'relief' phenomenon upon metallographic polishing. The eutectic regions consist of intermetallic compounds (IMC) as indicated by arrows in Figure 1, which are dispersed within Sn-rich matrix. The bright IMC particles, about 0.10–0.70 μm in size are identified as Ag_3Sn (in the form of thin platelets) whereas grey IMC particles, about 0.40–2.90 μm, are identified as Cu_6Sn_5, according to EDX analysis. These particles are well documented in the literature [18,19] and therefore can be identified from morphology visible using Secondary electron imaging under the SEM as shown in top-view (Figure 1a) as well as cross-section view (Figure 1b,c).

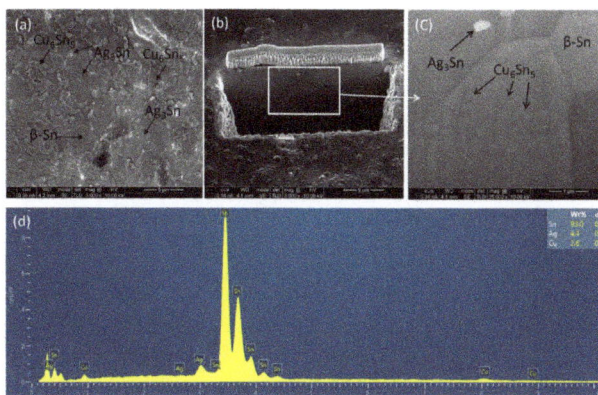

Figure 1. Sn–Ag–Cu solder alloy: (**a**) top-view SEM micrograph; (**b**) cross-section view SEM micrograph; (**c**) zoom-in view of the area marked in (**b**) and (**d**) energy dispersive X-ray (EDX) spectra at (**a**) with quantitative analysis as an insert.

These IMCs provide the strength of matrix material (β-Sn) which is relatively softer and has low elastic modulus and yield strength [20]. Thus, the role of such IMCs is somewhat similar to that of reinforcing particles in metal matrix composites [21]. Therefore, the presence of high volume fraction of β-Sn results in reduced elastic modulus and yield strength for the alloy. In contrast, the presence of IMCs increases elastic modulus and yield strength and provides stiffness of the alloy for structural applications. Towards better performance of solder alloys in practical applications, both of these properties are foreseen and hereby, a perfect balance between the volume fractions of such hard and soft faces are prerequisite. As the IMCs in as cast SAC alloy is not sufficient enough to provide the strength of the alloy [22], reinforcing particles in the form of either Fe or Al_2O_3 was added in it as mentioned in experimental section. The microstructure together with EDX analysis of SAC1 alloy is shown in Figure 2.

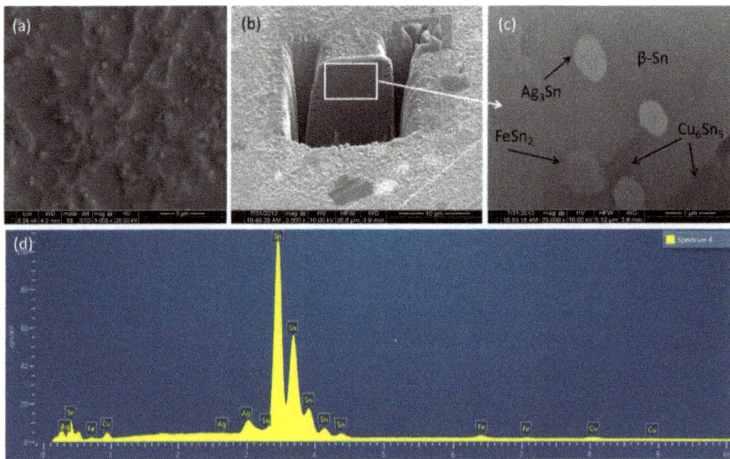

Figure 2. Sn–Ag–Cu solder alloy reinforced with Fe nano-particles (SAC1): (**a**) top-view SEM micrograph; (**b**) cross-section view SEM micrograph; (**c**) zoom-in view of the area marked in (**b**) and (**d**) EDX spectra at (**a**).

Similar to SAC, the matrix is large primary β-Sn grains and IMCs are dispersed in the matrix. In addition to the IMCs seen in SAC, SAC1 also contain $FeSn_2$. Due to the limited solubility of Fe in β-Sn matrix, most of the Fe precipitates as $FeSn_2$ IMC or other forms, such as pure Fe in eutectic regions, as will be evident in XRD spectra presented in later sections. As reported in literature, Fe incorporated SAC alloys exhibits relatively larger primary β-Sn grains as a result of the large degree of undercooling than that of SAC [22]. However, it has been reported that the addition of Fe suppresses the coarsening of IMCs mainly due to small amount of Fe in Ag_3Sn IMCs and thus helps towards grain refinement. The microstructure together with EDX analysis of SAC2 alloy is shown in Figure 3.

Similar to SAC and SAC1, the matrix is large primary β-Sn grains and IMCs are dispersed in the matrix. Distribution of Al_2O_3 nanoparticles in the matrix is uniform with a preferential trend to be allocated along β-Sn grain boundaries (Figure 3a). Examination of Cu_6Sn_5 IMCs by SEM shows that there is no significant change in size and distribution of Cu_6Sn_5 IMCs between composite and monolithic samples. It suggests that, volume percentage of reinforcement particles is not sufficient to significantly affect the growth dynamics of the Cu_6Sn_5 in the case of composite samples.

It is a well understood that, fine particles in alloys effectively restrict dislocation movement and thus provide higher yield strength. However, these particles may grow in size due to aging effect, in service and thus loss their impingement properties. In addition to that, when the coherency of the particles within the matrix gradually lost with particle growth, yield strength decreased further [23].

As Al_2O_3 nanoparticles are hard and thermally stable at relatively high temperature, compared to the IMCs form in the microstructure, it is expected that Al_2O_3 nanoparticles will restrict such grain growth during ageing/reflowing more effectively than that of $FeSn_2$ and other IMCs. Detailed understanding of such behavior as the strength of the alloys and effect of aging on strength is out of the scope of the present paper and accounted for in our future correspondences.

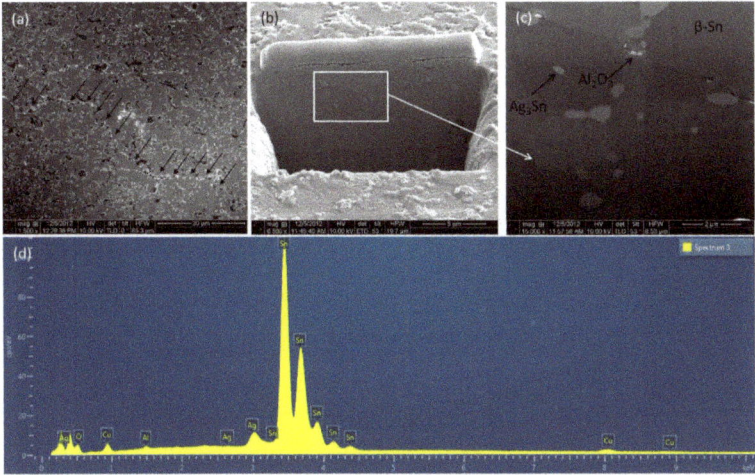

Figure 3. Sn–Ag–Cu solder alloy reinforced with Al_2O_3 nano-particles (SAC2): (**a**) top-view SEM micrograph; (**b**) cross-section view SEM micrograph; (**c**) zoom-in view of the area marked in (**b**) and (**d**) EDX spectra at (**a**).

3.2. Different Phases of Solder Alloys

The XRD spectra of SAC and nanocomposite solder alloys are shown in Figure 4. The scattering planes were (111) and (200) for Cu and (101) and (202) for Cu_6Sn_5. It is evident that, all the alloys including nanocomposites contains Cu_6Sn_5 and Ag_3Sn IMCs predominantly. In addition to that, SAC1 shows the presence of Fe_2Sn IMC. Due to high thermal stability, Al_2O_3 nanoparticles do not contribute towards any phase formation, however, retained as it is, as confirmed by SEM microstructural investigation reported in earlier Section 3.1 [24].

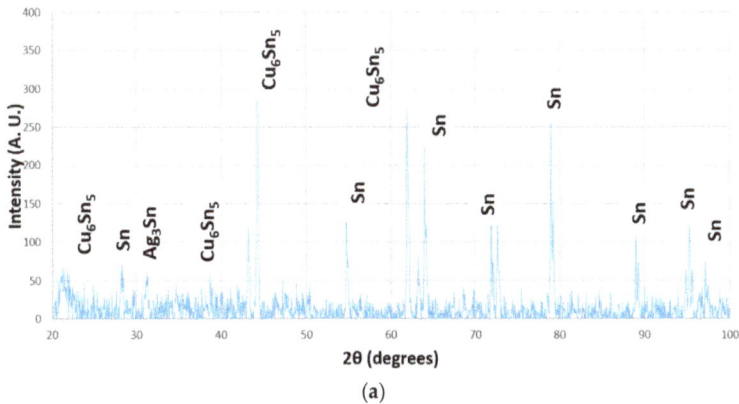

(**a**)

Figure 4. *Cont.*

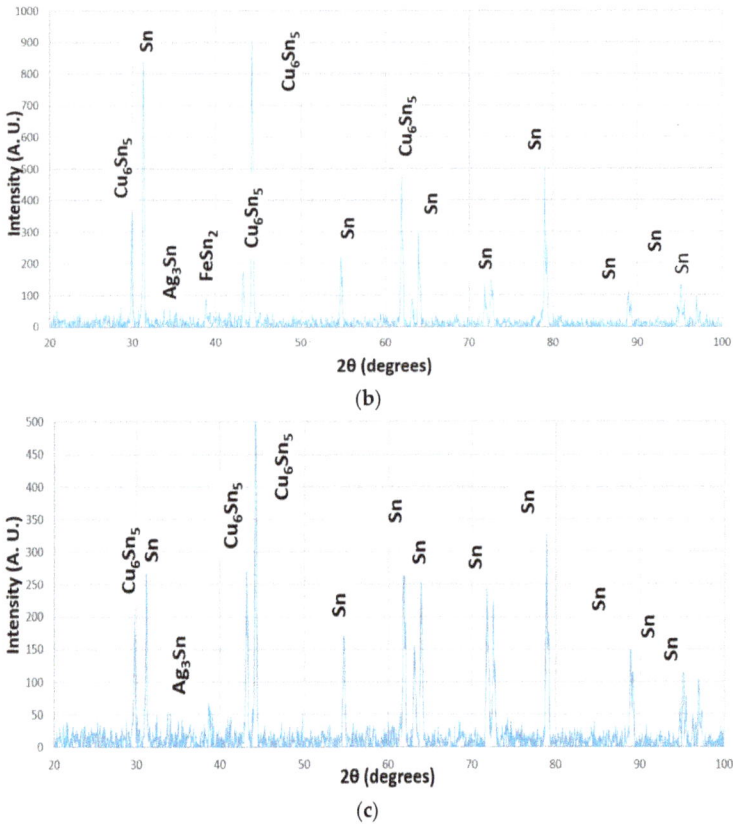

Figure 4. XRD spectra of solder alloys: (**a**) Sn–Ag–Cu alloy; (**b**) Sn–Ag–Cu alloy reinforced with Fe nanoparticles and (**c**) Sn–Ag–Cu alloy reinforced with Al$_2$O$_3$ nano-particles.

3.3. Thermal Analysis of Solder Alloys

Thermal analysis of the samples was carried out in order to determine the effect of nanoparticles addition in monolithic SAC alloy as shown in Figures 5 and 6. Figure 5 shows thermal gravimetric analysis (TGA) of solder alloys over temperature ranges from 50 °C to 400 °C. As evident, there was not any significantly change over the temperature range which suggest high thermal stability of these material up to 400 °C. Figure 6 shows the differential scanning calorimetry (DSC) of the alloys. Melting temperature of monolithic SAC solder alloy vary with the addition of different nanoparticles (Al$_2$O$_3$, C and Fe) as shown in Figure 6 from DSC analysis. However, this temperature difference took place over a very narrow range, i.e., between 220 °C and 230 °C. This confirms that the addition of nanoparticles in monolithic SAC solder alloy retain the melting temperature in desired range, which is suitable for their intended applications [25,26]. The area under DSC peaks represent energy density of solder alloys and it is 176.02, 175.5 and 178.01 J/g for SAC, SAC1 and SAC2 alloy, respectively. Thus, these materials have the potential to store thermal energy during their phase change from solid to liquid. As shown in Figure 6, melting temperatures are 220.9 °C and 220.8 °C for SAC1 and SAC2 as compared with 221.4 °C for SAC. Thus, with the addition of Fe or Al$_2$O$_3$, melting temperatures only changed slightly with respect to SAC alloy. Fe-bearing solders (SAC1) solidify at relatively lower temperatures than that of SAC during cooling as difficult nucleation of β-Sn can results in high degrees of undercooling prior to solidification. This indicates that β-Sn phase requires a large degree of undercooling to nucleate and solidify, explaining the existence of relatively larger primary β-Sn grains in SAC1 as reported in Section 3.1. The onset transformation temperature for exothermic descent of the curves, which represents the onset of melting, does not change markedly. It is understood that by adding Fe or Al_2O_3 in SAC, alloy composition has moved away from eutectic values, as shown by the effect of melting point peak broadening from DSC results (Figure 6) [25,26].

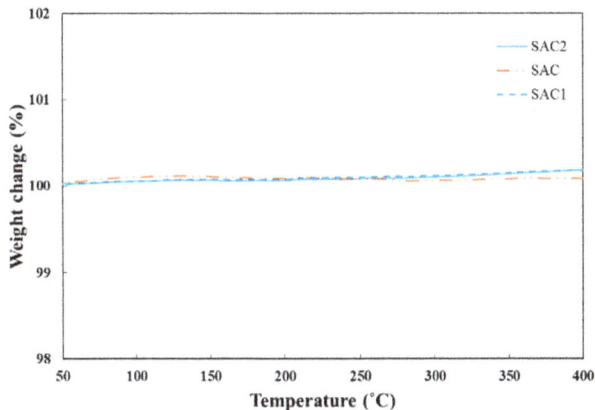

Figure 5. TGA analysis of solder alloys.

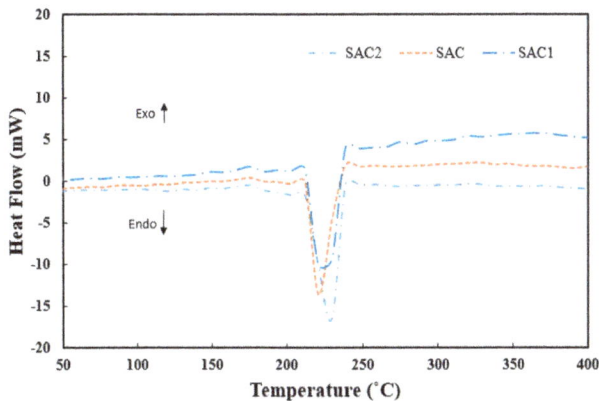

Figure 6. Differential scanning calorimetry (DSC) analysis of solder alloys.

The present investigation towards the development Pb-free solder alloys reports the physical and thermal characteristics of mentioned alloys. In this context, investigation on strength and mechanical properties of such alloys are foreseen in our future communications.

4. Conclusions

This current research reports the effect of small amount Fe or Al_2O_3 nanoparticles addition on microstructure and thermal behaviors of monolithic Pb-free Sn–Ag–Cu alloy. Based on experimental outcomes, the following conclusions can be made:

1. Fe-bearing solder nanocomposite form relatively larger primary β-Sn grains compared to monolithic Sn–Ag–Cu alloy. Fe addition also helps to form $FeSn_2$ IMCs dispersed in matrix and foreseen to restrict grain/ IMCs growth during aging/reflowing.

2. Addition of Al_2O_3 nanoparticles refine β-Sn grain size, dispersed in the matrix with preferential trend to be accumulated along grain boundaries.

3. Neither Fe nor Al_2O_3 nanoparticle addition cause any significant effect on thermal behavior compared to Sn–Ag–Cu solder alloys.

Author Contributions: Animesh K. Basak and Alokesh Pramanik conceived and designed the experiments; Hamidreza Riazi and Mahyar Silakhori performed XRD and thermal analysis experiments; Animesh K. Basak, Alokesh Pramanik and Angus K. O. Netting wrote the paper.

Conflicts of Interest: The authors declare no conflict of interest.

References

1. Miller, C.M.; Anderson, I.E.; Smith, J.F. A Viable Tin-Lead Solder Substitute: Sn-Ag-Cu. *J. Electron. Mater.* **1994**, *23*, 595–662. [CrossRef]

2. Anderson, I.E.; Cook, B.A.; Harringa, J.L.; Terpstra, R.L. Sn-Ag-Cu Solders and Solder Joints: Alloy Development, Microstructure, and Properties. *JOM* **2002**, 26–29. [CrossRef]

3. Liu, C.Z.; Chen, J. Nanoindentation of lead-free solders in microelectronic packaging. *Mater. Sci. Eng. A* **2007**, *448*, 340–344. [CrossRef]

4. Zou, H.F.; Yang, H.J.; Zhang, Z.F. Coarsening mechanisms, texture evolution and size distribution of Cu_6Sn_5 between Cu and Sn-based solders. *Mater. Chem. Phys.* **2011**, *131*, 190–198. [CrossRef]

5. Deng, X.; Chawla, N.; Chawla, K.K.; Koopman, M. Deformation behaviour of (Cu, Ag)—Sn intermetallics by nanoindentation. *Acta Mater.* **2004**, *52*, 4291–4303. [CrossRef]

6. Gao, F.; Takemoto, T. Mechanical properties evolution of Sn-3.5Ag based lead-free solders by nanoindentation. *Mater. Lett.* **2006**, *60*, 2315–2318. [CrossRef]

7. Gebhardt, E.; Petzow, G. Ueber den Auf-bau des Systems Silber-Kupfer-Zinn. *Zeitschrift fuer Metallkunde* **1959**, *50*, 597–605.

8. Loomans, M.E.; Fine, M.E. Tin-silver-copper eutectic temperature and composition Metall. *Mater. Trans. A* **2000**, *31A*, 1155–1162. [CrossRef]

9. Moon, K.W.; Boettinger, W.J.; Kattner, U.R.; Biancaniello, F.S.; Handwerker, C.A. Experimental and thermodynamic assessment of Sn-Ag-Cu solder alloys. *J. Electron. Mater.* **2000**, *29*, 1122–1136. [CrossRef]

10. Pramanik, A.; Basak, A.K.; Dong, Y.; Shankar, S.; Littlefair, G. Milling of nanoparticles reinforced Al-based metal matrix composites. *J. Compos. Sci.* **2018**, *2*, 13. [CrossRef]

11. Chellvarajoo, S.; Abdullah, M.Z. Microstructure and mechanical properties of Pb-free Sn–3.0Ag–0.5Cu solder pastes added with NiO nanoparticles after reflow soldering process. *Mater. Des.* **2016**, *90*, 499–507. [CrossRef]

12. Zhong, X.L.; Gupta, M. Development of lead-free Sn—0.7Cu/Al_2O_3 nanocomposite solders with superior strength. *J. Phys. D Appl. Phys.* **2008**, *41*, 095403. [CrossRef]

13. Salleh, M.A.A.M.; McDonald, S.D.; Nogita, K. Effects of Ni and TiO_2 additions in as-reflowed and annealed Sn0.7Cu solders on Cu substrates. *J. Mater. Process. Technol.* **2017**, *242*, 235–245. [CrossRef]

14. Niranjani, V.L.; Rao, B.S.S.C.; Sarkar, R.; Kamat, S.V. The influence of addition of nano sized molybdenum and nickel particles on creep behavior of Sn–Ag lead free solder alloy. *J. Alloys Comp.* **2012**, *542*, 136–141. [CrossRef]

15. Tao, Q.B.; Benabou, L.; Le, V.N.; Hwang, H.; Lu, D.B. Viscoplastic characterization and post-rupture microanalysis of a novel lead-free solder with small additions of Bi, Sb and Ni. *J. Alloys Comp.* **2017**, *694*, 892–904. [CrossRef]

16. Wu, C.M.L.; Yu, D.Q.; Law, C.M.T.; Wang, L. Properties of lead-free solder alloys with rare earth element additions. *Mater. Sci. Eng. R* **2004**, *44*, 1–44. [CrossRef]

17. Matin, M.A.; Vellinga, W.P.; Geers, M.G.D. Microstructure evolution in a Pb-free solder alloy during mechanical fatigue. *Mater. Sci. Eng. A* **2006**, *431*, 166–174. [CrossRef]

18. Rao, B.S.S.C.; Weng, J.; Shen, L.; Lee, T.K.; Zeng, K.Y. Morphology and mechanical properties of intermetallic compounds in SnAgCu solder joints. *Microelectron. Eng.* **2010**, *87*, 2416–2422.

19. Kumar, V.; Fang, Z.Z.; Liang, J.; Dariavach, N. Microstructural Analysis of Lead-Free Solder Alloys. *Metall. Mater. Trans. A* **2006**, *37*, 2505–2514. [CrossRef]

J. Compos. Sci. **2018**, *2*, 28

20. Deng, X.; Koopman, M.; Chawla, N.; Chawla, K.K. Young's modulus of (Cu, Ag)–Sn intermetallics measured by nanoindentation. *Mater. Sci. Eng. A* **2004**, *364*, 240–243. [CrossRef]

21. Basak, A.K.; Pramanik, A.; Islam, M.N. Failure mechanisms of nanoparticle reinforced metal matrix composite. *Adv. Mater. Res.* **2013**, *774*, 548–551. [CrossRef]

22. Shnawah, D.A.A.; Said, S.B.M.; Sabri, M.F.M.; Badruddin, I.A.; Che, F.X. Microstructure, mechanical, and thermal properties of the Sn-1Ag-0.5Cu solder alloy bearing Fe for electronics applications. *Mater. Sci. Eng. A* **2012**, *551*, 160–168. [CrossRef]

23. Dieter, G.E. *Mechanical Metallurgy*, 2nd ed.; McGraw-Hill: Minato-ku, Tokyo, 1976.

24. Basak, A.K.; Eddine, W.Z.; Celis, J.P.; Matteazzi, P. Characterisation and tribological investigation on thermally processed nanostructured Fe-based and Cu-based cermet materials. *J. Nanosci. Nanotechnol.* **2010**, *10*, 1179–1184. [CrossRef] [PubMed]

25. Elmer, J.W.; Specht, E.D.; Kumar, M. Microstructure and In Situ Observations of Undercooling for Nucleation of β-Sn Relevant to Lead-Free Solder Alloys. *J. Elelctron. Mater.* **2010**, *39*, 273–282. [CrossRef]

26. Reid, M.; Punch, J.; Collins, M.; Ryan, C. Effect of Ag content on the microstructure of Sn-Ag-Cu based solder alloys. *Solder. Surf. Mt. Technol.* **2008**, *20*, 3–8. [CrossRef]

© 2018 by the authors. Licensee MDPI, Basel, Switzerland. This article is an open access article distributed under the terms and conditions of the Creative Commons Attribution (CC BY) license (http://creativecommons.org/licenses/by/4.0/).

Journal of
composites science

MDPI

Article

Polylactic Acid Reinforced with Mixed Cellulose and Chitin Nanofibers—Effect of Mixture Ratio on the Mechanical Properties of Composites

Antonio Norio Nakagaito [1], Sohtaro Kanzawa [2] and Hitoshi Takagi [1,*]

[1] Graduate School of Technology, Industrial and Social Sciences, Tokushima University,
    Tokushima 770-8506, Japan; nakagaito@tokushima-u.ac.jp
[2] Graduate School of Advanced Technology and Science, Tokushima University, Tokushima 770-8506, Japan;
    chat.n0ir.66.lululu@gmail.com
* Correspondence: takagi@tokushima-u.ac.jp; Tel.: +81-88-656-7359

Received: 2 May 2018; Accepted: 15 June 2018; Published: 19 June 2018

Abstract: The development of all-bio-based composites is one of the relevant aspects of pursuing a carbon-neutral economy. This study aims to explore the possibility to reinforce polylactic acid by the combination of cellulose and chitin nanofibers instead of a single reinforcement phase. Polylactic acid colloidal suspension, cellulose and chitin nanofiber suspensions were mixed using only water as mixing medium and subsequently dewatered to form paper-like sheets. Sheets were hot pressed to melt the polylactic acid and form nanocomposites. The combination of cellulose and chitin nanofiber composites delivered higher tensile properties than its counterparts reinforced with cellulose or chitin nanofibers alone. Cellulose and chitin appear to complement each other from the aspect of the formation of a rigid cellulose nanofiber percolated network, and chitin acting as a compatibilizer between hydrophobic polylactic acid and hydrophilic cellulose.

Keywords: cellulose; chitin; nanofiber; polylactic acid; paper; compression molding

---

1. Introduction

Due to pressing ecological issues of modern civilization, the need to find new substitute materials that minimize environmental footprint is becoming ever more urgent. A substantial part of materials used in everyday life consist of polymers because they are lightweight, versatile, low cost, and easy to manufacture. However, polymers and derived composites mostly aim at long term durability to the detriment of easy disposability. They are generally made from fossil-based synthetic materials, but polymers are also produced by plants and animals through biochemical reactions. These naturally synthesized polymers are known as biopolymers. Among these biomass-derived polymers, cellulose is the most abundant polysaccharide comprising 40% of the organic matter on earth [1]. Cellulose is mostly found in the cell wall of plant fibers, as the structural reinforcement that provides the mechanical rigidity to support the plants' bodies. This framework is comprised of tiny semi-crystalline fibrous elements known as cellulose nanofibers, possessing mechanical properties similar to those of aramid fibers. The Young's modulus of the crystalline portions were measured to be 138 GPa [2], whereas the estimated tensile strength along the length of the nanofiber is in the range of 1.6 to 3 GPa [3].

Chitin is another highly plentiful polysaccharide available in nature, also found in the form of nanofibers in the exoskeleton of marine crustaceans, insects and in various filamentous fungi. The molecular structure of chitin is identical to that of cellulose, apart from the fact that a hydroxyl group of every glucose ring is replaced by an acetamido group [1]. The reduced number of hydroxyl groups makes chitin less hydrophilic than cellulose.

Polylactic acid is the first commodity large-volume biopolymer available. It is a polyester synthesized by condensation polymerization of lactic acid, a naturally occurring organic compound that can be obtained by fermentation of sugar or starch-derived feedstocks. Polylactic acid, from here on abbreviated as PLA, is a versatile thermoplastic that can be processed in conventional polymer processing equipment into films, fibers, and injection-molded parts. Besides, under appropriate composting conditions PLA can be easily decomposed into carbon dioxide and water.

Numerous composites of PLA reinforced with either cellulose or chitin nanofibers have been developed by various methods like film casting, melt compounding, and papermaking. Casting is the easiest way to obtain small samples in laboratory, through slow evaporation of water from aqueous suspensions of latex resin and nanofibers. Among the more industrially-oriented processes, the melt compounding method has been extensively studied by Oksman and coworkers. They started using microcrystalline cellulose, and after finding out that separation into whiskers did not occur during compounding [4], N,N-dimethylacetamide and lithium chloride mixture was used to swell the microcrystalline cellulose. Subsequent disintegration into whiskers while compounding with PLA in a twin screw extruder was accomplished with addition of polyethylene glycol to lower the viscosity of the compound [5]. However, the mechanical properties of composites were not improved over the neat PLA due to thermal degradation of cellulose and residual swelling agents and the presence of polyethylene glycol. Other processing aids like glycerol triacetate to disperse cellulose nanofibers [6], triethyl citrate to disperse chitin and/or cellulose nanofibers [7,8] produced composites with better mechanical properties compared to the PLA-plasticizer mixture, but still worse than the neat PLA. Kiziltas et al. tried poly hydroxyl butyrate as a disperser of cellulose nanofibers with similarly limited results [9]. Iwatake et al. came up with a different approach by dissolving PLA in an organic solvent and mixing it with cellulose nanofibers previously suspended in the same solvent before melt compounding [10], and the study was continued by Suryanegara et al. with promising results [11]. Oksman and colleagues, in turn, prepared high-cellulose nanofiber content master batch following a similar protocol and further diluting it with additional PLA by compounding in a twin-screw extruder [12]. The tensile modulus and strength were increased by over 20% relative to neat PLA at 5 wt % nanofiber load. Acetylation to hydrophobize cellulose nanofibers and subsequent melt compounding from master batch did not show significant difference from non-treated nanofibers though [13]. A similar study was reported by Li et al. [14] who used polyethylene glycol and polyethylene oxide to enhance dispersion of chitin nanofibers with PLA powder in aqueous suspension, and freeze-drying prior to melt compounding. However, the aqueous mixture of only nanofibers and PLA gave the best mechanical properties, confirming the negative effect of adding other substances. In effect, the tensile stress increased with nanofiber content up to 30 wt % [15]. Another of their studies proposed the use of sodium ionomer to enhance flowability of the molten compound as viscosity increases with nanofiber content [16]. Flexural properties more than tripled over neat PLA at a nanofiber load of 40 wt %. Impact toughness was increased by about 300% as well. Even though melt compounding is an established industrial process, the mechanical properties are limited due to difficulties in nanofiber dispersion especially at high contents, and to the lack of the formation of a stiff percolated network of nanofibers linked by hydrogen bonds [17–20]. The percolation phenomenon has been long recognized as the reason ordinary papers are made by cellulosic pulp fibers mutually adhered by hydrogen bonds, without the need for adhesives. The importance of percolation on nano-scale cellulose reinforcements was first noticed by Favier et al. [21,22] in 1995. Nanocomposites made by film casting of cellulose whiskers and latex resin showed shear modulus at rubbery state staying constant up to the temperature of cellulose decomposition while the modulus of the pure matrix resin had a decreasing slope. The phenomenon was attributed to a percolation effect of the cellulose whiskers, forming a stiff framework connected by hydrogen bonds. Studies using chitin whiskers showed varied results, in some cases producing partial formation of percolated networks [23] and in others the formation of a rigid network [24]. Later, the use of cellulose nanofibers also confirmed the occurrence of percolation [25]. Following studies however, identified an important difference in the sense that whiskers formed networks linked solely by hydrogen bonds whereas nanofibers formed networks by hydrogen bonds and mechanical entanglements due to their flexible nature as opposed to stiff whiskers [26–30]. Due to its high hydrophilicity, the most effective way to achieve good dispersion of cellulose nanofibers in hydrophobic PLA is by mixing them in aqueous medium. Although PLA is insoluble in water, aqueous suspensions can be obtained by using PLA short fibers or particles, that can be easily mixed with cellulose nanofibers in water suspension. After dewatering, the mixture forms paper-like sheets that can be laminated and compression molded. Previous attempts by the paper-making method were effective in reinforcing PLA with cellulose nanofibers [31–33] and chitin nanofibers [34]. Our previous studies concerning nanocomposites of PLA with cellulose compared to those with chitin nanofibers showed that both nanofiber types can reinforce the matrix up to high nanofiber contents. Cellulose delivered higher reinforcement at high nanofiber loads whereas chitin showed better reinforcement at lower nanofiber loads. In composites containing less reinforcing nanofibers, the percolated network of microfibrillated cellulose (MFC) would be weaker while the chitin nanofiber (ChNF)-PLA interaction would deliver the mechanical strength, favoring the ChNF-rich composites. Composites with higher contents of nanofibers are expected to be stronger at MFC-rich compositions as the MFC percolation is predominant. Therefore, the natural following step would be to evaluate the effect of both nanofibers acting together as the reinforcing phase. However, studies on the combination of both cellulose and chitin nanofibers as reinforcements have been scarce, or even unavailable especially when the matrix is PLA. The aim of this study was to explore the possibility of joint reinforcement by cellulose and chitin nanofibers of PLA matrix, and confirm the validity of this approach. The experiments demonstrated that it is possible to produce cellulose-chitin hybrid nanocomposites with mechanical properties superior to those of nanocomposites reinforced solely with cellulose or chitin alone. The combination of cellulose and chitin nanofibers as reinforcing phase opens the opportunity to develop nanocomposites with better mechanical performances and perhaps helps to lower the cost of raw materials. Chitin nanofibers demand less energy intensive nanofibrillation treatments to be produced than cellulose nanofibers [35].

2. Materials and Methods

2.1. Materials

The polylactic acid used as matrix phase comprised of the aqueous colloidal suspension types Landy PL-1000, Landy PL-2000, and Landy PL-3000 produced by Miyoshi Oil and Fat Co., Ltd., Tokyo, Japan. All varieties have weak anionic character and consist of 40 wt % PLA particles suspended in dispersing agents. Average particle sizes are 5, 2, and 1 μm for Landy PL-1000, PL-2000, and PL3000, respectively. The original industrial applications were as highly heat-resistant adhesive (PL-1000), coating and thermal adhesive (PL-2000), and coating (PL-3000). Cellulose nanofibers consisted of a commercially available microfibrillated cellulose (MFC) morphology of tradename Celish KY-100G provided by Daicel Corporation, Tokyo, Japan. The chitin nanofiber (ChNF) was extracted from purified chitin powder from crab shells (Nacalai Tesque, Inc., Kyoto, Japan) by grinding, following a previously reported protocol [34] that delivered fibrils with diameters below 100 nm. First, 30 g of chitin powder was suspended in 2 L of distilled water in which 30 g of acetic acid was added and stirred for 16 h. The acidic medium protonates the amino portion of the acetoamido groups of chitin, producing electrostatic repulsion that facilitates individualization of nanofibers. Subsequently, the powder aqueous suspension was passed through an ultra-fine friction grinder Supermasscolloider MKCA6-2 (Masuko Sangyo Co., Ltd., Saitama, Japan). With the grindstone aperture adjusted to a point when the stones slightly touch each other at a rotational speed of 1500 rpm, the suspension was poured into the grinder and the aperture was immediately closed by 0.15 mm. A tiny aperture is maintained by the hydrodynamic pressure created by the wet spinning grindstone. The suspension was passed twice through the grinder while 1 L of distilled water was gradually added to avoid evaporation and drying up of the grindstone aperture that would stop rotation. The obtained suspension after fibrillation therefore totaled 3 L, with a ChNF concentration of 1 wt %.

2.2. Nanocomposite Fabrication

Thin nanocomposites with the same reinforcing phase content of 50 wt % but varying ChNF to MFC ratios of 1:0, 4:1, 3:2, 1:1, 2:3, 1:4, and 0:1 were fabricated. The necessary amount (0–1.5 g) of MFC was diluted in 300 g of distilled water and stirred for one hour. Next, the ChNF aqueous suspension was slowly added by a dropper to the suspension so that the amount of added ChNF (0–1.5 g) with MFC totaled 3 g. The ChNF-MFC suspension was diluted by adding distilled water up to 600 g and the stirring was continued for another hour. The PLA Landy PL-2000 weighing 3 g (colloidal suspension containing 40 wt % PLA particles) was diluted in 200 g of distilled water and stirred for 30 minutes. This PLA suspension was slowly dripped to the ChNF-MFC suspension and further stirred for three hours. Finally, the suspension was stirred under reduced pressure for at least one hour, in order to eliminate entrapped air bubbles. The obtained suspensions were vacuum filtered through Buchner funnel and filter paper 110 mm in diameter Advantec 101 (Toyo Roshi Kaisha, Ltd., Tokyo, Japan). Retentates were peeled off from the filter papers, sandwiched in between filter papers Advantec 2 (Toyo Roshi Kaisha, Ltd., Tokyo, Japan) and perforated metal plates and oven-dried at 50 °C for 48 h. Dried sheets were further dried at 105 °C for one hour to completely remove moisture, and compression molded at 180 °C and 10 MPa for five minutes.

The same protocol was adopted to produce nanocomposites with the same ChNF to MFC ratios, but with amounts of reinforcing and matrix phases adjusted to achieve total reinforcement contents of 25 wt % and 75 wt %. For these composites, the matrix consisted of a mixture of Landy PL-1000 and Landy PL-3000 in a 1:1 ratio, since the variety Landy PL-2000 production was halted during the course of this study.

Additional thicker specimens were fabricated for Izod impact strength test and heat deflection temperature measurement. Nanocomposites with reinforcing phase content of 50 wt % with different ChNF to MFC ratios of 4:1, 3:2, 1:1, 2:3, 1:4, and 0:1 were prepared. The ChNF-MFC-PLA aqueous suspensions were prepared following the already described method and using Landy PL-1000 and Landy PL-3000 1:1 mixture. Suspensions were vacuum filtered through Buchner funnel and filter paper Advantec 101 (Toyo Roshi Kaisha, Ltd.) with 270 mm in diameter. Since the retentates were too thin, an additional step was necessary to mold thicker pieces. The retentates were released from the filter paper and put inside a circular form 115 mm in diameter with a meshed bottom covered with filter paper and molded by placing a 3.5 kg weight on top of it for one hour. Next the molded cake was oven dried at 70 °C for 48 h and at 105 °C for another 24 h. Dried cakes were cut into 10 mm by 80 mm pieces, further dried at 105 °C for one hour and compression molded at 190 °C and 10 MPa for 20 min.

2.3. Tensile Test

Specimens with dimensions of 10 mm by 80 mm were subjected to tensile test using an Instron 5567 (Instron Corp., Norwood, MA, USA) universal materials testing machine equipped with a 5 kN load cell, at a strain rate of 1 mm/min and the gage length set to 30 mm. To prevent damage at the gripping points, the ends of each specimen had carton paper tabs glued at both sides and clasped with serrated chucks. As the specimens were ribbon shaped and the failure would occur at different places, the widths and thicknesses were measured at equally spaced points along the length. The resulting cross sectional areas corresponding to the actual fracture sites measured before fracture were considered to calculate the tensile modulus and strength.

2.4. Heat Deflection Temperature (HDT)

The deflection temperature of nanocomposites was measured following the Japanese Industrial Standards JIS K7191-3 under a load of 0.45 MPa (Method B) and 1.8 MPa (Method A). Specimen dimensions were 80 mm in length, 10 mm in width, and 4 mm in thickness.

2.5. Izod Impact Strength Test

Impact resistance of nanocomposites was measured by an Izod impact tester (No. 158, Yasuda Seiki Seisakusho, Ltd., Hyogo, Japan). Notched specimens 80 ± 2 mm long, 4.0 ± 0.2 mm thick, 10.0 ± 0.2 mm wide with remaining width of 8.0 ± 0.2 mm at the notch, were tested according to the Japanese Industrial Standards JIS K7110.

2.6. X-ray Computed Tomography

The 3D imaging of samples were obtained by a high resolution desk-top micro-CT SKY Scan 1172 (Bruker-Micro CT, Kontich, Belgium).

3. Results and Discussion

The results of tensile test of the thin ChNF-MFC-PLA nanocomposites are shown in Figure 1. It depicts nanocomposites containing 50 wt % nanofibers but with differing ChNF to MFC ratios. Among the nanocomposites with varying ChNF to MFC compositions, there is a clear predominance of the ChNF-MFC mixed nanocomposites over those reinforced solely either with ChNF or MFC. In this 50 wt % nanofiber load case, the equal amounts of chitin and cellulose (ChNF to MFC ratio of 1:1) produced the best tensile property results. Although it seems counterintuitive, the phenomenon might be explained in terms of the number of hydroxyl groups present in chitin and cellulose. Cellulose molecule has three hydroxyl groups attached to each glucose ring, whereas chitin retains only two as the C-2 position is occupied by an acetamido group, making chitin less hydrophilic than cellulose. At the higher MFC content end of the ChNF-MFC composition spectrum, the percolation of cellulose nanofibers forms a stiff network interconnected by hydrogen bonds that confers much of the strength to the composites. On the other extreme containing more chitin, the more hydrophobic character relative to cellulose makes chitin more compatible with the hydrophobic PLA, with the percolated network playing a lesser role on strength. However, when chitin and cellulose were combined, both percolation of cellulose and better affinity of chitin with the PLA matrix seem to have worked in concert to enhance the mechanical properties of the composites. This hypothesis is based on the results of a previous study [34] in which ChNF-PLA composites were stronger than MFC-PLA composites at lower nanofiber contents, while MFC-PLA composites were stronger at higher nanofiber loadings. In order to verify this assumption, nanocomposites with different reinforcing nanofiber contents were later produced and tested. For lesser reinforcing nanofiber containing composites, the percolated network of MFC would be weaker while the ChNF-PLA interaction would be dominant, resulting in a shift of the mechanical property peak to the ChNF-rich side of composites. On the other hand, composites containing higher amounts of nanofibers would have the peak shifted towards the MFC-rich composites due to the MFC percolation controlling the mechanical properties.

At the time of fabrication of a new batch of composites, the production of the colloidal suspension type Landy PL-2000 had been discontinued by the manufacturer. Unable to purchase the same product, other two varieties Landy PL-1000 and Landy PL-3000 still in production were purchased and mixed in a 1 to 1 ratio, to replace PL-2000. The mixture was not intended to serve as an equivalent substitute for Landy PL-2000, as the mechanical properties of the composites with the new PLA matrix decreased relative to the previous one. However, the measurements are still valid for comparative purposes concerning variations in ChNF to MFC ratio. The results of tensile test are presented in Figure 2. As predicted, the composites containing 75 wt % nanofibers had the peak of strength (ChNF to MFC ratio of 2:3) shifted towards higher MFC content. This means that at higher nanofiber content, the contribution of MFC to the mechanical properties of composites is dominant, by the formation of a percolated network that promotes stiffness and strength. Looking at the composites with low nanofiber content of 25 wt % in Figure 3, even though it is subtle, the highest strength is displaced to the side with higher ChNF content. In this case, the percolated framework of MFC becomes less relevant to the strength of composites, and the affinity of ChNF with PLA matrix becomes determinant to the properties of the composite. Once again, the combination of ChNF and MFC reinforcing nanofibers delivered higher strength than ChNF or MFC reinforcements alone. The data depicted in Figures 1–3 are summarized in Figure 4.

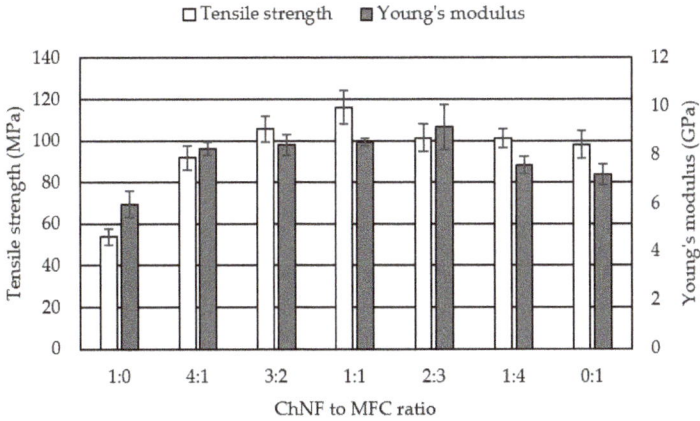

Figure 1. Tensile strength and Young's modulus of nanocomposites with 50 wt % reinforcing phase with varying chitin nanofiber (ChNF) to microfibrillated cellulose (MFC) ratios.

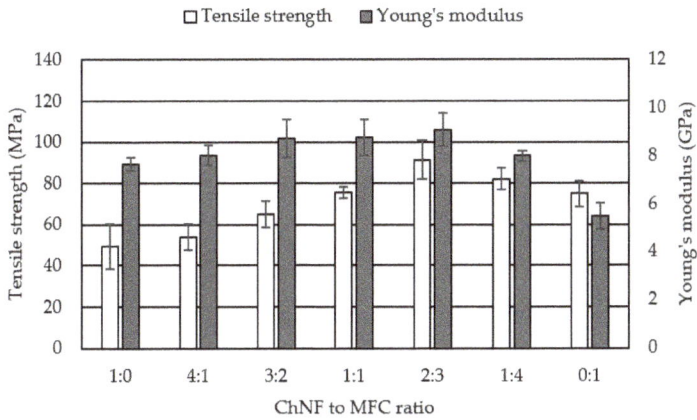

Figure 2. Tensile strength and Young's modulus of nanocomposites with 75 wt % reinforcing phase with varying chitin nanofiber (ChNF) to microfibrillated cellulose (MFC) ratios.

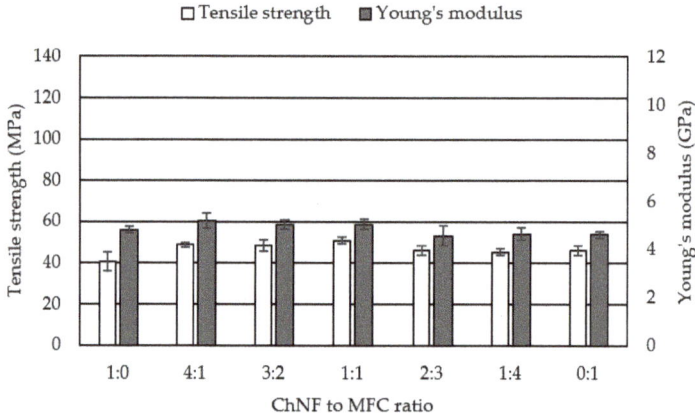

Figure 3. Tensile strength and Young's modulus of nanocomposites with 25 wt % reinforcing phase with varying chitin nanofiber (ChNF) to microfibrillated cellulose (MFC) ratios.

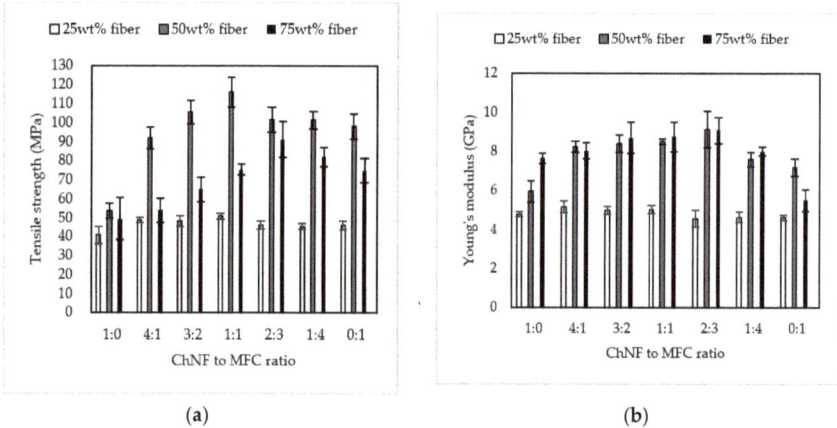

(a)	(b)

Figure 4. Plots summarizing the data of all composites (reinforcement fraction of 25 wt %, 50 wt %, and 75 wt %) as a function of varying chitin nanofiber (ChNF) to microfibrillated cellulose (MFC) ratios: (a) tensile strength; (b) tensile modulus.

To evaluate the mechanical properties of composites at higher temperatures, the heat deflection temperature (HDT) was measured. As shown in Figure 5, by Method B of HDT measurement, all nanocomposites warped in the direction opposite to the application of the load of 0.45 MPa as the temperature was increased. This phenomenon was likely caused by a ChNF-MFC distribution gradient along the thickness of the specimens. Due to the fabrication process, reinforcing fibrils tend to concentrate on the bottom side of the specimens during dewatering, effectively restraining the thermal expansion of the PLA matrix at one side of the composite. As the temperature approached 200 °C, the only nanocomposite to deflect by 0.34 mm was PLA containing solely MFC, with an HDT of 191 °C. As the remaining nanocomposites did not show any deflection up to 200 °C, they were subjected to a higher load of 1.8 MPa following Method A of measurement, and the results are depicted in Figure 6. The HDT for nanocomposites containing ChNF to MFC ratios of 1:4, 1:1, and 4:1 were 60, 75, and 96 °C respectively, increasing with increments in ChNF content. This is attributed to the intermediate hydrophilic ChNF that delivered a better bond between the hydrophobic PLA and hydrophilic MFC, delivering good stress transfer even at high temperature.

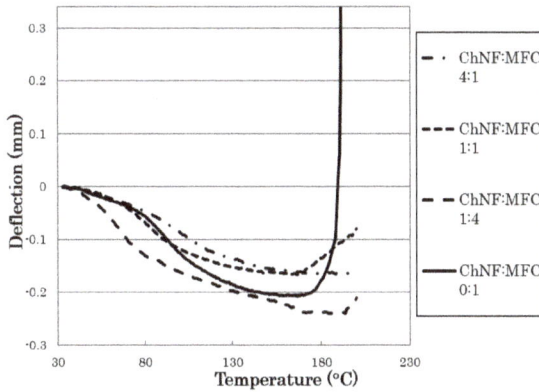

Figure 5. Heat deflection temperature (HDT) curves by the Method B (load of 0.45 MPa) of nanocomposites with 50 wt % reinforcing phase with varying chitin nanofiber (ChNF) to microfibrillated cellulose (MFC) ratios.

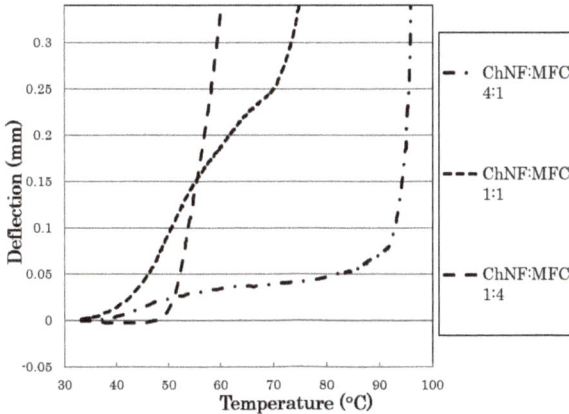

Figure 6. Heat deflection temperature (HDT) curves by the Method A (load of 1.8 MPa) of nanocomposites with 50 wt % reinforcing phase with varying chitin nanofiber (ChNF) to microfibrillated cellulose (MFC) ratios.

Izod impact test results are shown in Figure 7. The ChNF-MFC nanocomposites containing 50 wt % nanofibers delivered impact resistance higher than neat PLA and similar to that of MFC-reinforced composite, with the ChNF to MFC ratio 3:2 composite showing slightly higher mean impact strength. However, considering the whole series of measured values and the size of the error intervals, there were no significant differences in impact resistance when chitin and cellulose were combined. Although a clear change was not observed when reinforced with ChNF-MFC mixture or MFC only, it is an indication that the combination of cellulose and chitin nanofibers had no negative effect on the impact resistance of composites, corroborating the results of enhanced impact strength obtained in chitin nanofiber-reinforced PLA reported by Li et al. [14]. According to their reasoning, the impact strength is obtained owing to uniformly dispersed nanofibers forming a network that absorbs large amounts of energy during the fracture. The state of nanofibers dispersion of the composites in the present study is shown by X-ray tomography images in Figure 8. It is difficult to state that dispersion was achieved at nano-scale since the resolution is limited to a few micrometers, but agglomerations at micro-scale can be observed. There are apparent differences in dispersion depending on the ChNF to MFC ratio, but they did not significantly affect the impact resistance of the nanocomposites.

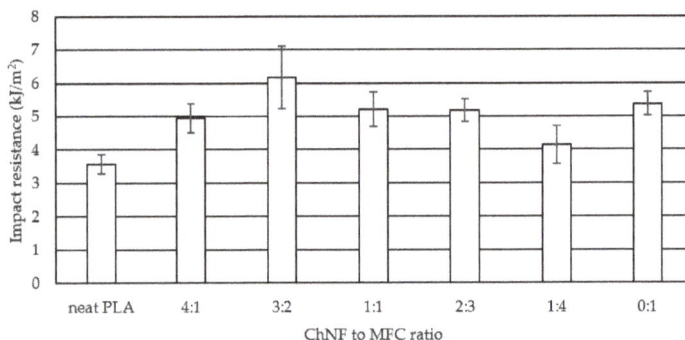

Figure 7. Notched impact resistance of nanocomposites with 50 wt % reinforcing phase with varying chitin nanofiber (ChNF) to microfibrillated cellulose (MFC) ratios.

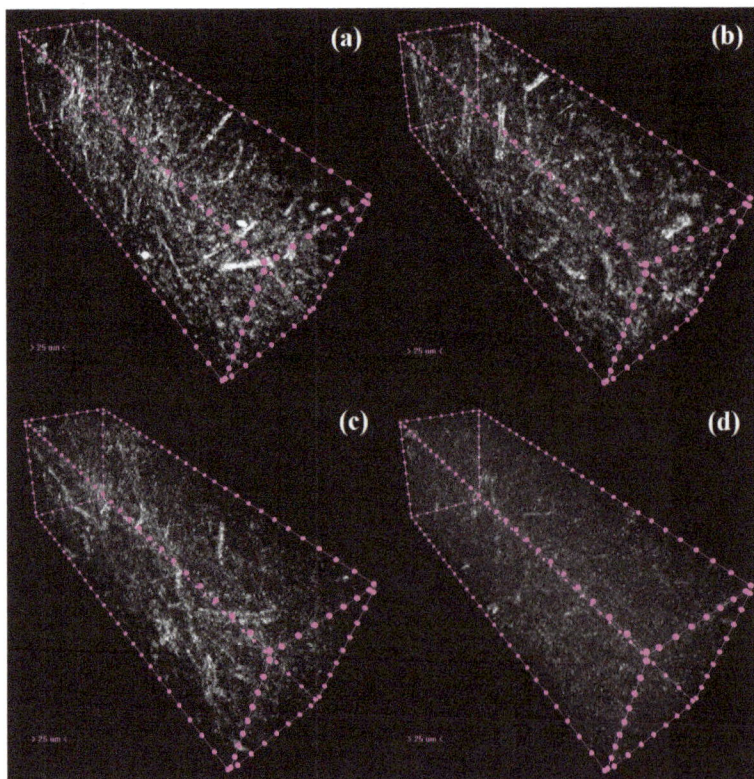

Figure 8. X-ray computed tomography of nanocomposites with 50 wt % reinforcing phase with varying chitin nanofiber (ChNF) to microfibrillated cellulose (MFC) ratios: (**a**) 4:1; (**b**) 1:1; (**c**) 1:4; (**d**) 0:1 (all MFC-reinforced). Due to the resolution limitation, the ChNF comprised of sub-micrometer diameter elements is not observable, whereas the portion of MFC larger than a few micrometers is seen as white spots.

Considering all the mechanical properties evaluated for the nanocomposites, it was revealed that the combination of ChNF and MFC to reinforce PLA produces higher mechanical performance than PLA reinforced with ChNF or MFC separately. Even though each type of reinforcing nanofiber provides significant reinforcement to PLA, there is some kind of synergistic effect that enhances the reinforcement when mixed. The combination of hydrophilic MFC, less hydrophilic ChNF and hydrophobic PLA composite resembles the composition of plant fibers. The cell wall of plant fibers is a biocomposite made of a framework of cellulose embedded in hydrophobic lignin matrix, in addition to compounds called hemicelluloses. Hemicelluloses have a hydrophilic character in between cellulose and lignin, working as a compatibilizer to hydrophilic cellulose and hydrophobic lignin. In that sense, the ChNF-MFC-PLA can be seen as a biomimetic nanocomposite system. This study revealed that not only is it possible to reinforce hydrophobic resins with either cellulose or chitin nanofibers, but the combination of these nanofibers can further improve the reinforcing effect. Even though the extraction of cellulose nanofibers is difficult and costly, the extraction of chitin nanofibers is relatively easier and therefore less costly [35]. The proper combination of these two nanofibers may bring a method of reducing the production cost of cellulose-based nanocomposites while enhancing their mechanical properties.

4. Conclusions

This study assessed the effect of the combination of chitin and cellulose nanofibers on the reinforcement of polylactic acid composites fabricated by a papermaking process. The following conclusions can be summarized.

1. The tensile strength, tensile modulus, and heat deflection temperature of chitin-cellulose nanocomposites were increased relative to their counterparts reinforced only by chitin nanofibers or only by cellulose nanofibers;
2. The reinforcing mechanism of chitin-cellulose nanofibers is presently not fully understood and requires additional studies;
3. The chitin-cellulose reinforcement increases the impact resistance of PLA with values on a par with cellulose nanofiber-reinforced PLA;
4. The overall cost of nanocomposites could potentially be reduced by addition of chitin nanofibers as they are easier to extract than cellulose nanofibers.

Author Contributions: A.N.N. and H.T. conceived and designed the experiments; S.K. performed the experiments; A.N.N., S.K., and H.T. analyzed the data; A.N.N. wrote the paper.

Funding: This research received no external funding.

Acknowledgments: Authors are indebted to T. Semba from Kyoto Municipal Institute of Industrial Technology and Culture, Polymer Material Lab., Kyoto, Japan, for heat deflection temperature (HDT) measurements and K. Abe from Research Institute for Sustainable Humanosphere, Kyoto University, Japan, for the X-ray CT imaging.

Conflicts of Interest: The authors declare no conflict of interest.

References

1. Stevens, E.S. *Green Plastics: An Introduction to the New Science of Biodegradable Plastics*, 1st ed.; Princeton University Press: Princeton, NJ, USA, 2002; pp. 83–103.
2. Nishino, T.; Takano, K.; Nakamae, K. Elastic-modulus of the crystalline regions of cellulose polymorphs. *J. Polym. Sci. Part B Polym. Phys.* **1995**, *33*, 1647–1651. [CrossRef]
3. Saito, T.; Kuramae, R.; Wohlert, J.; Berglund, L.A.; Isogai, A. An Ultrastrong nanofibrillar biomaterial: the strength of single cellulose nanofibrils revealed via sonication-induced fragmentation. *Biomacromolecules* **2013**, *14*, 248–253. [CrossRef] [PubMed]
4. Mathew, A.P.; Oksman, K.; Sain, M. Mechanical properties of biodegradable composites from poly lactic acid (PLA) and microcrystalline cellulose (MCC). *J. Appl. Polym. Sci.* **2005**, *97*, 2014–2025. [CrossRef]

5.   Oksman, K.; Mathew, A.P.; Bondeson, D.; Kvien, I. Manufacturing process of cellulose whiskers/polylactic acid nanocomposites. *Compos. Sci. Technol.* **2006**, *66*, 2776–2784. [CrossRef]

6.   Herrera, N.; Mathew, A.P.; Oksman, K. Plasticized polylactic acid/cellulose nanocomposites prepared using melt-extrusion and liquid feeding: Mechanical, thermal and optical properties. *Compos. Sci. Technol.* **2015**, *106*, 149–155. [CrossRef]

7.   Herrera, N.; Singh, A.A.; Salaberria, A.M.; Labidi, J.; Mathew, A.P.; Oksman, K. Triethyl citrate (TEC) as a dispersing aid in polylactic acid/chitin nanocomposites prepared via liquid-assisted extrusion. *Polymers* **2017**, *9*, 406. [CrossRef]

8.   Herrera, N.; Salaberria, A.M.; Mathew, A.P.; Oksman, K. Plasticized polylactic acid nanocomposite films with cellulose and chitin nanocrystals prepared using extrusion and compression molding with two cooling rates: Effects on mechanical, thermal and optical properties. *Compos. Part A Appl. Sci. Manuf.* **2016**, *83*, 89–97. [CrossRef]

9.   Kiziltas, A.; Nazari, B.; Erbas Kiziltas, E.; Gardner, D.J.; Han, Y.; Rushing, T.S. Method to reinforce polylactic acid with cellulose nanofibers via a polyhydroxybutyrate carrier system. *Carbohydr. Polym.* **2016**, *140*, 393–399. [CrossRef] [PubMed]

10.   Iwatake, A.; Nogi, M.; Yano, H. Cellulose nanofiber-reinforced polylactic acid. *Compos. Sci. Technol.* **2008**, *68*, 2103–2106. [CrossRef]

11.   Suryanegara, L.; Nakagaito, A.N.; Yano, H. The effect of crystallization of PLA on the thermal and mechanical properties of microfibrillated cellulose-reinforced PLA composites. *Compos. Sci. Technol.* **2009**, *69*, 1187–1192. [CrossRef]

12.   Jonoobi, M.; Harun, J.; Mathew, A.P.; Oksman, K. Mechanical properties of cellulose nanofiber (CNF) reinforced polylactic acid (PLA) prepared by twin screw extrusion. *Compos. Sci. Technol.* **2010**, *70*, 1742–1747. [CrossRef]

13.   Jonoobi, M.; Mathew, A.P.; Abdi, M.M.; Makinejad, M.D.; Oksman, K. A Comparison of modified and unmodified cellulose nanofiber reinforced polylactic acid (PLA) prepared by twin screw extrusion. *J. Polym. Environ.* **2012**, *20*, 991–997. [CrossRef]

14.   Li, J.; Gao, Y.; Zhao, J.; Sun, J.; Li, D. Homogeneous dispersion of chitin nanofibers in polylactic acid with different pretreatment methods. *Cellulose* **2017**, *24*, 1705–1715. [CrossRef]

15.   Li, J.; Li, J.; Feng, D.; Zhao, J.; Sun, J.; Li, D. Comparative study on properties of polylactic acid nanocomposites with cellulose and chitin nanofibers extracted from different raw materials. *J. Nanomater.* **2017**, *2017*. [CrossRef]

16.   Li, J.; Li, J.; Feng, D.; Zhao, J.; Sun, J.; Li, D. Excellent rheological performance and impact toughness of cellulose nanofibers/PLA/ionomer composite. *RSC Adv.* **2017**, *7*, 28889–28897. [CrossRef]

17.   Nair, K.G.; Dufresne, A. Crab shell chitin whisker reinforced natural rubber nanocomposites. 2. Mechanical behavior. *Biomacromolecules* **2003**, *4*, 666–674. [CrossRef] [PubMed]

18.   Nair, K.G.; Dufresne, A. Crab shell chitin whisker reinforced natural rubber nanocomposites. 1. Processing and swelling behavior. *Biomacromolecules* **2003**, *4*, 657–665. [CrossRef] [PubMed]

19.   Dalmas, F.; Cavaille, J.Y.; Gauthier, C.; Chazeau, L.; Dendievel, R. Viscoelastic behavior and electrical properties of flexible nanofiber filled polymer nanocomposites. Influence of processing conditions. *Compos. Sci. Technol.* **2007**, *67*, 829–839. [CrossRef]

20.   Dalmas, F.; Chazeau, L.; Gauthier, C.; Cavaille, J.Y.; Dendievel, R. Large deformation mechanical behavior of flexible nanofiber filled polymer nanocomposites. *Polymer* **2006**, *47*, 2802–2812. [CrossRef]

21.   Favier, V.; Canova, G.R.; Cavaille, J.Y.; Chanzy, H.; Dufresne, A.; Gauthier, C. Nanocomposite materials from latex and cellulose whiskers. *Polym. Adv. Technol.* **1995**, *6*, 351–355. [CrossRef]

22.   Favier, V.; Chanzy, H.; Cavaille, J.Y. Polymer nanocomposites reinforced by cellulose whiskers. *Macromolecules* **1995**, *28*, 6365–6367. [CrossRef]

23.   Morin, A.; Dufresne, A. Nanocomposites of chitin whiskers from Riftia tubes and poly(caprolactone). *Macromolecules* **2002**, *35*, 2190–2199. [CrossRef]

24.   Paillet, M.; Dufresne, A. Chitin whisker reinforced thermoplastic nanocomposites. *Macromolecules* **2001**, *34*, 6527–6530. [CrossRef]

25.   Dufresne, A.; Cavaille, J.Y.; Vignon, M.R. Mechanical behavior of sheets prepared from sugar beet cellulose microfibrils. *J. Appl. Polym. Sci.* **1997**, *64*, 1185–1194. [CrossRef]

26. Dufresne, A.; Dupeyre, D.; Vignon, M.R. Cellulose microfibrils from potato tuber cells: Processing and characterization of starch-cellulose microfibril composites. *J. Appl. Polym. Sci.* **2000**, *76*, 2080–2092. [CrossRef]
27. Dufresne, A.; Vignon, M.R. Improvement of starch film performances using cellulose microfibrils. *Macromolecules* **1998**, *31*, 2693–2696. [CrossRef]
28. Angles, M.N.; Dufresne, A. Plasticized starch/tunicin whiskers nanocomposite materials. 2. Mechanical behavior. *Macromolecules* **2001**, *34*, 2921–2931. [CrossRef]
29. Angles, M.N.; Dufresne, A. Plasticized starch/tunicin whiskers nanocomposites. 1. Structural analysis. *Macromolecules* **2000**, *33*, 8344–8353. [CrossRef]
30. Samir, M.A.S.A.; Alloin, F.; Paillet, M.; Dufresne, A. Tangling effect in fibrillated cellulose reinforced nanocomposites. *Macromolecules* **2004**, *37*, 4313–4316. [CrossRef]
31. Nakagaito, A.N.; Fujimura, A.; Sakai, T.; Hama, Y.; Yano, H. Production of microfibrillated cellulose (MFC)-reinforced polylactic acid (PLA) nanocomposites from sheets obtained by a papermaking-like process. *Compos. Sci. Technol.* **2009**, *69*, 1293–1297. [CrossRef]
32. Larsson, K.; Berglund, L.A.; Ankerfors, M.; Lindström, T. Polylactide latex/nanofibrillated cellulose bionanocomposites of high nanofibrillated cellulose content and nanopaper network structure prepared by a papermaking route. *J. Appl. Polym. Sci.* **2012**, *125*, 2460–2466. [CrossRef]
33. Wang, T.; Drzal, L.T. Cellulose-nanofiber-reinforced poly(lactic acid) composites prepared by a water-based approach. *ACS Appl. Mater. Interfaces* **2012**, *4*, 5079–5085. [CrossRef] [PubMed]
34. Nakagaito, A.N.; Yamada, K.; Ifuku, S.; Morimoto, M.; Saimoto, H. Fabrication of chitin nanofiber-reinforced polylactic acid nanocomposites by an environmentally friendly process. *J. Biobased Mater. Bioenergy* **2013**, *7*, 152–156. [CrossRef]
35. Fan, Y.M.; Saito, T.; Isogai, A. Preparation of chitin nanofibers from squid pen beta-chitin by simple mechanical treatment under acid conditions. *Biomacromolecules* **2008**, *9*, 1919–1923. [CrossRef] [PubMed]

© 2018 by the authors. Licensee MDPI, Basel, Switzerland. This article is an open access article distributed under the terms and conditions of the Creative Commons Attribution (CC BY) license (http://creativecommons.org/licenses/by/4.0/).

Journal of
composites science

MDPI

Article

The Effect of Polycaprolactone Nanofibers on the Dynamic and Impact Behavior of Glass Fibre Reinforced Polymer Composites

Cristobal Garcia [1,*], Irina Trendafilova [1] and Andrea Zucchelli [2]

[1] Department of Mechanical and Aerospace Engineering, University of Strathclyde, 75 Montrose Street, Glasgow G1 1XJ, UK; irina.trendafilova@strath.ac.uk
[2] Department of Industrial Engineering, University of Bologna, Viale Risorgimento 2, 40125 Bologna, Italy; a.zucchelli@unibo.it
* Correspondence: cristobal.garcia@strath.ac.uk; Tel.: +44-141-548-5025

Received: 13 June 2018; Accepted: 18 July 2018; Published: 23 July 2018

Abstract: In this article, the effect of polycaprolactone nanofibers on the dynamic behavior of glass fiber reinforced polymer composites is investigated. The vibratory behavior of composite beams in their pristine state (without any nano modification) and the same beams modified with polycaprolactone fibers is considered experimentally. The experimental results show that the incorporation of polycaprolactone nanofibers increases the damping; however, it does not significantly affect the natural frequencies. Additionally, the paper analyses the effect of polycaprolactone nanofibers on the impact behavior of glass fiber/epoxy composites. This has already been analyzed experimentally in a previous study. In this work, we developed a finite element model to simulate the impact behavior of such composite laminates. Our results confirm the conclusions done experimentally and prove that composites reinforced with polycaprolactone nanofibers are more resistant to damage and experience less damage when subjected to the same impact as the pristine composites. This study contributes to the knowledge about the dynamic behavior and the impact resistance of glass fiber reinforced polymer composites reinforced with polycaprolactone nanofibers. The findings of this study show that interleaving with polycaprolactone nanofibers can be used to control the vibrations and improve the impact damage resistance of structures made of composite mats as aircrafts or wind turbines.

Keywords: composite materials; nano composites; dynamic behavior; impact behavior; finite element model; electrospinning

1. Introduction

Composite laminates reinforced with polymeric electrospun nanofibers are attracting increasing attention among the scientific community due to their superior material properties. It is well known that the incorporation of electrospun nanofibers into the interfaces of composite laminates can drastically change/improve some material properties. For example, the inclusion of polycaprolactone nanofibers in the interfaces of composite laminates can be used to increase the mode I fracture toughness up to 50% [1]. The introduction of tetraethyl orthosilicate electrospun nanofibers in the epoxy resin of glass fiber composites was found to enhance the interlaminar shear strength up to 15% [2]. Core-shell polyamide nanofibers can be used to prepare flame-retardant polymer nanofibers [3], which can be potentially used to develop composite laminates with enhanced flame retardancy. Thus, electrospun nanofibers offer great potential to improve some structural properties of composite mats [4,5].

In the last years, the dynamic properties of composite laminates have been widely studied because of their applications in delamination detection [6] and structural health monitoring [7]. However, only several studies have investigated the vibratory behavior of composites reinforced with electrospun nanofibers. For example, the authors of [8] demonstrated that the incorporation of nylon nanofibers in glass fiber composite mats increases the damping ratio; however, it does not considerably affect the natural frequencies. Similar results were found in [9], where an important increase of the damping ratio in carbon fiber composite laminates due to the incorporation of nylon nanofibers is found. To date, the dynamic and vibratory behavior of composite laminates reinforced with electrospun nanofibers has been poorly investigated. Therefore, there is an urgent need to explore the effect of electrospun nanofibers (e.g., polycaprolactone or polyamide) on the vibration properties of composite laminates.

In the last decade, several papers have investigated the effect of electrospun nanofibers on the mechanical properties of composite laminates. For example, the authors of [10] investigated the effect of nylon nanofibers on the interlaminar properties of glass fiber/epoxy laminates. The results reveal that the addition of nylon nanofibers increased the mode I and mode II energy releases rates by 62% and 109%, respectively. Other works such as [11] reported that the maximum stress of carbon fiber composites is significantly enhanced (with an increment of 35%) due to the incorporation of nylon polymer nanofibers into the composite laminates. To date, most of the studies have investigated the effect of electrospun nanofibers on the fracture toughness in opening and sliding mode [12,13], on the interlaminar shear strength [14], on the tensile strength [15], and the compression strength after impact [16]. However, there are still very few works that have reported the effect of electrospun nanofibers on the impact behavior of composite laminates.

In this work, the authors experimentally investigate the influence of polycaprolactone nanofibers on the vibratory properties of glass fiber reinforced polymer composites. For that purpose, composite beams without nanofibers (which will be referred to as pristine) and with polycaprolactone nanofibers (referred to as nano) are subjected to free vibration tests. Subsequently, the vibration signals of the pristine and nano composites are acquired through an accelerometer. The signals are further analyzed to evaluate the natural frequencies and the damping ratio of the two types of specimens. The main goal is to assess the influence of polycaprolactone nanofibers on the natural frequencies and damping ratio of the glass fiber epoxy composite beams. To the best of our knowledge, this work is the first attempt to study the vibratory behavior of composites reinforced with polycaprolactone nanofibers.

The second part of the manuscript numerically analyzes the effect of polycaprolactone nanofibers on the impact behavior of glass fiber reinforced composites. For this study, a finite element model is used to evaluate the impact response of composites with and without polycaprolactone nanofibers. The numerical results obtained are compared to the experimental results published by [17] and they show quite good agreement regarding the impact damaged area of the pristine and the nano modified specimens. Both the experimental and the numerical results show that the incorporation of polycaprolactone nanofibers in the composite interfaces significantly enhances the impact damage resistance of the glass fiber composites. On the basis of these results, it can be concluded that composites reinforced with polycaprolactone nanofibers are less prone to impact damage than the pristine composites. As far as the authors are aware, this is the first time that a finite element model is used to simulate the impact damage resistance of composites reinforced with polycaprolactone nanofibers.

The main contributions of this study are twofold: Firstly, the paper demonstrates that the interleaving with polycaprolactone nanofibers can be used to reduce the composite vibrations, which have important applications for composite structures in which vibrations are a source of problems. Secondly, the paper reveals that the addition of polycaprolactone nanofibers can also be used to develop composite structures with a higher resistance to impacts (e.g., bird strikes or hailstones), which is important for the health of the composite mats used in aircrafts, wind turbines, and other civil structures.

The rest of the manuscript is organized as follows: The second section is devoted to the fabrication of the pristine and nano-modified composites. The third section explains the methodology used to assess the natural frequencies and the damping ratio of the composite beams. Section 4 presents the finite element model used to simulate the impact response of the composite specimens. Section 5 presents and discusses in detail the experimental and numerical results. The last section offers some conclusions.

2. Fabrication of Composites with and without Polycaprolactone Nanofibers

This section describes the fabrication of the glass fiber/epoxy composites reinforced with and without polycaprolactone nanofibers. Figure 1 illustrates the composite lay-up of the pristine and nano-modified composites used in this study. Pristine composites (without polycaprolactone nanofibers) were fabricated by hand lay-up of eight layers of unidirectional glass fiber epoxy prepreg, as detailed in Figure 1a. The composite specimens are prepared with dimensions of 168 mm × 32 mm × 3.1 mm and stacking sequence $[0,90,0,90]_s$. After the lay-up, the composite beams are cured using a vacuum bag in an autoclave at 150° for about one hour, as indicated in the supplier's specifications. The weight fractions of the glass fiber and epoxy resin are 78.6% and 21.4%, respectively, for the pristine composites.

The nano composites (with polycaprolactone nanofibers) are also manufactured by hand lay-up with identical composite prepregs, number of layers, dimensions, ply orientations, and curing process as the pristine composites. However, six layers of polycaprolactone nanofibers are interleaved at each of the composite interfaces (excluding the central one), as shown in Figure 1b. It is also important to mention that the effect of the nanofibers on the final thickness of the composites is negligible (less than 1%). Additionally, the difference in weight for the pristine and nano composites due to the incorporation of polycaprolactone nanofibers is also negligible. Therefore, the weight fraction of polycaprolactone nanofibers in the nano composites is very small (less than 1%).

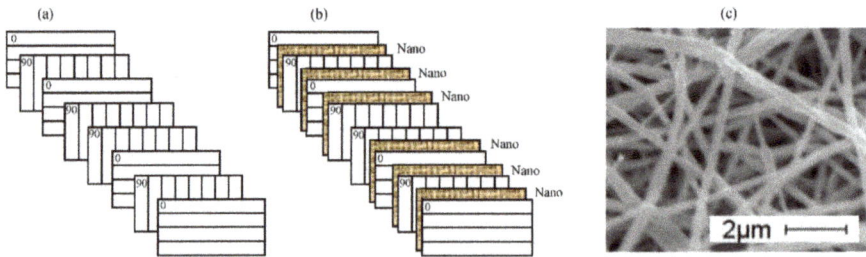

Figure 1. Composite lay-up of the (**a**) pristine and (**b**) nano modified composites. (**c**) Scanning electron microscope (SEM) image of polycaprolactone nanofibers as spun, adapted from figure in [17].

The layers of polycaprolactone nanofibers were prepared by the electrospinning technique [18]. This procedure was chosen because is an easy and low-cost technology to prepare polycaprolactone nanofibers with a wide variety of morphologies. For the preparation of the fibers, polycaprolactone pellets are dissolved in a solvent mixture of formic acid and acetic acid (60/40) at 15% w/v. Subsequently, the chemical solution is transferred to a syringe to be spun using the following operational conditions: a high voltage of 23 kV, a feed rate of 0.9 mL/h, and a needle tip-collector distance of 15 cm. As a result, an ultrathin layer of interconnected polycaprolactone nanofibers is obtained, as depicted in Figure 1c. The scanning electron microscope (SEM) image is adapted from [17] and shows a dense array of polycaprolactone nanofibers distributed randomly in the membrane. The diameter of the polycaprolactone nanofibers is 275 nm, with a standard deviation of 75 nm.

3. Vibratory Behavior of Pristine and Polycaprolactone Nano-Modified Beams

The aim of this section is to describe the vibration tests used to obtain the natural frequencies and the damping ratio of the pristine and polycaprolactone nano-modified composites. The experimental procedure used to measure the natural frequencies and the damping is given in Figure 2a and can be divided into four steps. First, the composite beams are fixed at both ends (clamped-clamped boundary conditions) using a test rig. The clamped regions are 8 mm long and the free span of the beams is 152 mm. Second, an impact excitation is applied to the middle of the composite beam using a modal hammer. Third, the free vibration response of the composite specimens is measured by an accelerometer (RT-440 portable analyzer, SKF, Gotemburgo, Sweden). The vibration signals are recorded for 0.5 s at a sampling rate of 5 kHz, as shown in Figure 2b, and each measurement is repeated ten times per specimen. Finally, the free vibration responses (Figure 2b) are used to calculate the first five natural frequencies using the fast Fourier transform [19] and the damping via the logarithm decrement method [20].

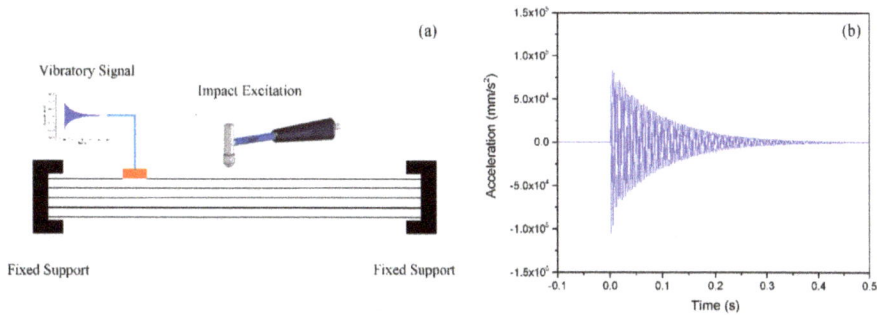

Figure 2. Schematic representation of the (**a**) vibration test used to measure the natural frequencies and damping in pristine and nano composites and (**b**) free vibration response of the composite specimens.

4. Impact Behavior of Pristine and Polycaprolactone Nano-Modified Beams

This section explains the methodology used to evaluate the impact response of the composites reinforced with and without polycaprolactone nanofibers. The section is organized as follows. The first part describes the finite element model used for studying the impact behavior of the composite specimens and the second part compares the results of the numerical model with available experimental data.

4.1. Numerical (FE) Modelling

This paragraph presents the finite element model used to simulate the impact damage resistance of the composite beams with and without polycaprolactone nanofibers. The composite laminates are modelled using ANSYS composite PrepPost. The layers of the composite laminates are modelled one by one including the ply thickness, stacking orientations, materials, and other heterogeneous features for each of the layers. The unidirectional glass fibre/epoxy layers are simulated using Solid 185 type elements and the material constants which are indicated in Table A1. The composite interfaces made of epoxy resin in the pristine composite and polycaprolactone nanofibers in the nano composite are modelled using cohesive elements.

Figure 3 represents the cohesive zone model used to simulate the initiation and the evolution of damage in the pristine and the nano-modified laminates. The cohesive model is based on the bilinear model proposed by Alfano and Crisfield [21]. From the figure, it can be deducted that the stiffness of the cohesive elements (K^0) is constant under small element displacements ($<\delta_{n,t}*$). However, the stiffness of the cohesive elements (K^1) decreases progressively when the level of displacement is above $\delta_{n,t}*$. The figure also shows that the initiation of the damage is defined by the displacement at maximum cohesive traction ($\delta_{n,t}*$). Therefore, for small displacements below $\delta_{n,t}*$, it is considered that there is no delamination/damage. As the material is non damaged, the stiffness is constant and equal to the original stiffness of the material. When the displacement of the elements is above $\delta_{n,t}*$, the damage progresses with the increase of the displacement $\delta_{n,t}$ and accordingly, the stiffness goes down with a factor of $(1 - D_m)$, as indicated in the equation shown in Figure 3.

Damage Initiation

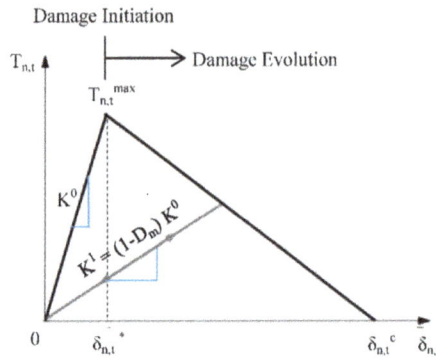

Figure 3. Schematic of the cohesive zone model used to simulate the composite interfaces.

The relation between the traction ($T_{n,t}$) and the displacement ($\delta_{n,t}$) of the cohesive elements can be defined using the following Equation (1).

$$\frac{T_{n,t}}{\delta_{n,t}} = (1 - D_m)K^{\circ}{}_{n,t} \tag{1}$$

where $K^{\circ}{}_{n,t}$ represents the initial stiffness of the cohesive element and $(1 - D_m)$ is a factor reduction of the stiffness due to the damage. The subscripts "n" and "t" refer to the normal and the tangential states. When the composites are undamaged, the damage parameter (D_m) is 0 and the value of the stiffness is equal to the original stiffness. When the composites are damaged, the D_m is between the two values 0 to 1 and the stiffness goes down, as indicated in Equation (1). Therefore, the level of damage is defined by the reduction of the stiffness. The value of $D_m = 1$ corresponds to complete debonding, which corresponds to the critical displacement $\delta_{n,t}{}^c$.

The corresponding cohesive parameters used to simulate the delamination between the composite interfaces for the pristine and the nano-modified composites are indicated in Table 1. From the table, it can be seen that the initial stiffness for the interfaces made of epoxy resin is 926 MPa/mm, while the stiffness for the interfaces of polycaprolactone nanofibers is 533 MPa/mm. The cohesive stiffness shows the same value for the normal and tangential components due to the isotropic nature of epoxy resin and the uniform distribution of polycaprolactone nanofibers in epoxy resin. The parameter alpha is defined as the ratio of $\delta_{n,t}{}^*$ to $\delta_{n,t}{}^c$ (see Figure 3) and can be used to calculate $\delta_{n,t}{}^*$. This is of the utmost importance as the area under the triangle 0, $\delta_{n,t}{}^*$, $T_{n,t}{}^{max}$ on Figure 3 corresponds to the energy needed to initiate delamination and the area under the triangle $\delta_{n,t}{}^*$, $\delta_{n,t}{}^c$, $T_{n,t}{}^{max}$ is associated with the energy for the damage propagation. The non-dimensional weighting parameter (betha) assigns different weights to the tangential and normal displacements, where we have assumed that the tangential and normal displacement contributes to the delamination (mixed mode debonding). The criteria used to select the material constants are shown in Appendix B.

Table 1. Cohesive zone parameters used to simulate the composite interfaces made of epoxy resin (pristine) and polycaprolactone nanofibers (nano).

Parameter	Abreviation	Pristine	Nano	Units
Maximum Normal Traction	$T_n{}^{max}$	5	2.8	MPa
Normal Displacement at Debonding	$\delta_n{}^c$	0.27	0.35	mm
Maximum Tangential Traction	$T_t{}^{max}$	5	2.8	MPa
Tangential Displacement at Debonding	$\delta_t{}^c$	0.27	0.35	mm
Ratio	α	0.02	0.015	Dimensionless
Non-Dimensional Parameter	β	1	1	Dimensionless
Initial Stiffness	$K^0{}_{n,t}$	926	533	MPa/mm

It has to be mentioned that in addition and together with the bilinear model proposed by Alfano and Crisfield [21], the Puck criterion is used when modelling damage initiation [22]. This criterion is based on the material properties, stresses, and strains, and incorporates the limit values of the strains in the different modes. According to this criterion, damage initiation is related to a certain value of stiffness reduction. In this study, the Puck criterion was used for modelling the interface delamination for the four failure modes of damage initiation related to the fibers and the matrix. These four modes are Tensile Fiber Failure Mode, Compressive Fiber Failure Mode, Tensile Matrix Failure Mode, and Compressive Matrix Failure Mode.

The damage evolution law utilized in the numerical simulations is based on the instant stiffness reduction. The stiffness reduction is used to define how the composite interfaces are degraded because of the damage. Accordingly, this stiffness reduction can vary between 0 and 1, where 0 indicates no reduction in the stiffness and 1 is associated with complete stiffness loss. In this study, we have assumed that there is an 80% reduction of stiffness reduction due to the delamination/damage for the four modes of damage.

4.2. Model Verification

To verify the results obtained using the finite element model and to validate the model, the finite element results are compared with the experimental results obtained in the paper of Saghafi et al. [17]. The numerical and the experimental results are presented in Table 2. The table represents the damaged area on the laminates with and without polycaprolactone nanofibers as a result of the impact with energies of 24 J and 36 J. From the table, it can be seen that the numerical results show the same trend as the reported experimental results and the delaminated area decreases for the nano modified samples for both cases of impact. Furthermore, it should be noted that the experimentally measured damaged area is in very good agreement with the damaged area obtained in the numerical simulations, which comes to further validate the results of the numerical simulations.

Table 2. Area damaged (mm^2) in pristine and nano composite due to the 24 and 36 J energy impact.

Energy	Sample	Experimental [17]	Numerical
24 J	Pristine	170 mm^2	175 mm^2
	Nano	125 mm^2	126 mm^2
36 J	Pristine	260 mm^2	275 mm^2
	Nano	197 mm^2	196 mm^2

The experimental results were conducted using a drop-weight impact machine as reported in [17]. For this purpose, the same composite specimens are impacted at 24 and 36 J using a drop-weight impact machine. The impactor consists of a steel spherical ball with a diameter and weight of 12.7 mm and 1.22 kg, respectively. Each impact was repeated three times for configuration.

5. Results and Discussion

This section analyzes the effect of polycaprolactone nanofibers on the dynamic and the impact behavior of glass fiber composite laminates. The section is divided into three parts. The first and second parts are devoted to the effect of polycaprolactone nanofibers on the natural frequencies and the damping, respectively. The last part investigates the effect of polycaprolactone nanofibers on the impact damage resistance of glass fiber composite laminates.

5.1. Effect of Polycaprolactone Nanofibers on the Natural Frequencies

The effect of polycaprolactone nanofibers on the natural frequencies of the composite laminates is analyzed in this paragraph. For that purpose, the first five natural frequencies of the composites with and without polycaprolactone nanofibers are determined using the results from the vibration test explained in Section 3. Table 3 shows the natural frequencies for the pristine and nano modified composites and the variation of the natural frequencies in percentage due to the incorporation of polycaprolactone nanofibers.

Table 3. Effect of polycaprolactone nanofibers on the natural frequencies.

Frequency (*f*)	Pristine (Hz)	Nano (Hz)	Variation (%)
First	484.5	461.5	4.7
Second	930.9	923.2	0.8
Third	1373.8	1369.8	0.3
Fourth	1857.5	1843.2	0.8
Fifth	2303.8	2278.0	1.2

The results reveal that the natural frequencies of the composites with polycaprolactone nanofibers are smaller compared to the pristine specimens. However, these changes are rather small and the variation of the natural frequencies is not significant (lower than 5%). In our view, the incorporation of nanofibers into the laminates induces little reduction in the composite stiffness, which slightly reduces the natural frequencies. In other words, the nanofibers create a matrix enrichment at the ply-to-ply interfaces, causing an increment of the matrix content, which reduces the composite stiffness and consequently the natural frequencies. These findings are in line with other results reported in the literature. For example, Ref. [23] reported that the natural frequencies of glass fiber composites showed minuscule changes (less than 4%) due to the inclusion of nylon nanofibers. Similar results were found in [24,25], which reported very small changes in the natural frequencies due to the inclusion of carbon nanofibers and nanotubes, respectively. In conclusion, it can be said that the changes of the natural frequencies depend on the properties of the nanofiber interleaved. For this particular case, the effect of the polycaprolactone nanofibers on the natural frequencies is miniscule and inconclusive as these variations are in the region of experimental and measurement error.

5.2. Effect of Polycaprolactone Nanofibers on the Damping Ratio

This paragraph analyses the effect of polycaprolactone nanofibers on the damping ratio of glass fiber composite laminates. The damping ratio for the laminates with and without nanofibers is calculated using the method described in Section 3. Table 4 illustrates the damping of the pristine and nano modified composites. Additionally, the table also includes the variation of damping ratio due to the reinforcement with nanofibers.

Table 4. Effect of polycaprolactone nanofibers on the damping ratio.

	Pristine (*Dimensionless*)	Nano (*Dimensionless*)	Variation (%)
Damping	0.01208	0.01277	5.7

The results indicate that the damping of the nano modified composites is higher with respect to the damping of the pristine panels. It can be seen that the damping ratio increased by 5.7% due to the reinforcement with polycaprolactone nanofibers. This can be attributed to the fact that nanofibers are able to dissipate energy, giving the nano modified composite a higher damping ratio. Thus, the nanofibrous mats play the role of dampers. Our results are in good agreement with other results published previously. For example, the authors from [8] suggest that the damping of glass fiber composites increased by 36% due to the interleaving with nylon nanofibers. Other works, such as [26,27], found that the damping of composite laminates increased up to 28% and 108% because of the interleaving with jute nanofibers and carbon nanotubes respectively. As a conclusion, it can be said that the inclusion of polycaprolactone nanofibers in the composite interfaces increases the damping ratio because the nanofibers act as dampers.

Therefore, composites interleaved with polycaprolactone nanofibers can be potentially used to reduce the amplitude of the vibrations in composite structures such as aircrafts, wind turbines, or bridges.

5.3. Effect of Polycaprolactone Nanofibers on the Impact Damage Behaviour

The aim of this section is to analyse the effect of polycaprolactone nanofibers on the impact damage resistance of glass fibre composites. For this study, the finite element model introduced in Section 4.1 is used to simulate the damage caused by impacts with energies of 24 J and 36 J in the pristine and nano modified composites.

The composite beams are subjected to impacts using energies of 24 J and 36 J, as detailed in Figure 4. The figure shows that the experimental and numerical force-displacement curves for the same energy impact are almost equal. Therefore, it can be concluded that the total impact energy used in the experiments and numerical simulations is the same. From the figure, the peak forces and displacements for the 24 and 36 J energy impacts can also be clearly observed. Therefore, according to our simulations, the peak force increases from 3795 to 4821 N when the energy of the impact varies between 24 and 36 J. The mechanical impacts are located at the center of the specimen and each impact is applied at the same location for each test. The composite beams are fixed using clamp-clamp boundary conditions. Therefore, the composite specimens are clamped using fixed supports at both ends of the composite beams.

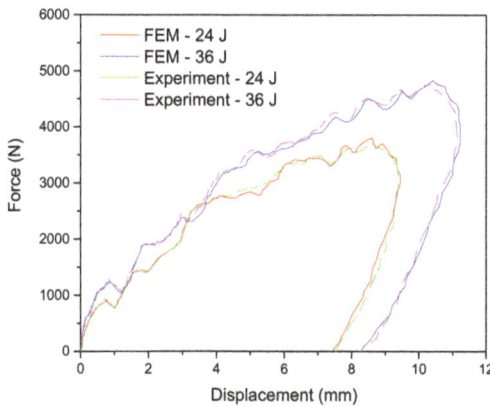

Figure 4. Comparison between the experimental and numerical force-displacement curves for the energy impacts at 24 J and 36 J.

The results obtained using the finite element model simulations are illustrated in Figure 5. The legend scale on the left refers to the level of damage in the composite specimen, where the strong blue colour (0%) represents the undamaged area of the composite and the red colour (100%) represents the severely damaged area of the composite. The green colour in between (50%) corresponds to damage states, which are in between the above two, the non-damaged and the severely damaged states, where the composite specimens are partially damaged. From the simulations, it can be clearly appreciated that nano composites are less damaged than the pristine panels for the two cases investigated (24 and 36 J). Therefore, it can be concluded that according to the finite element modelling and experiments, the incorporation of polycaprolactone nanofibers reduces the delaminated area by about 27% for the two energy levels at 24 and 36 J. This can be attributed to the good adhesion between polycaprolactone nanofibers and epoxy resin and the formation of heterogeneous separated phases on the composites interfaces, which increases the energy dissipation [28]. Additionally, the figure also shows that the most severe damage in the composite is located in the impacted zones (red area).

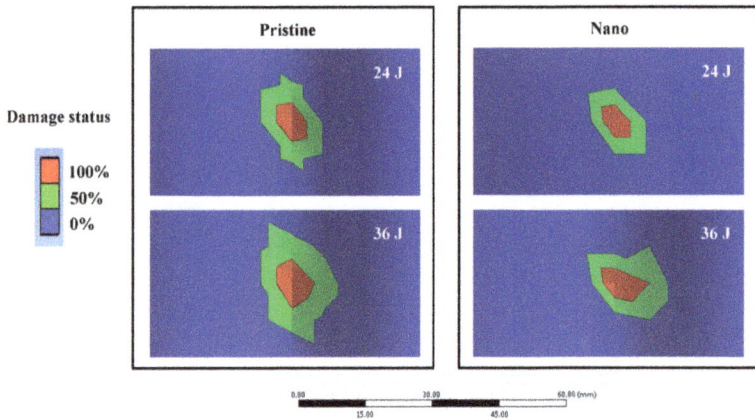

Figure 5. Area damaged after the energy impact at 24 and 36 J for the pristine and nano composites.

Experimental results show that the area damaged as a result of the 24 J impact is 170 and 125 mm^2 for the pristine and nano composites, respectively [17], and our numerical simulations confirm that the damaged area decreased from 175 to 126 mm^2 as a result of the incorporation of polycaprolactone nanofibers. The same trend is found for an impact with 36 J energy, where the area damaged in the composite laminates decreased by around 26% due to the reinforcement with polycaprolactone nanofibers. Therefore, the results obtained experimentally and numerically confirm that the composites reinforced with polycaprolactone nanofibers are less susceptible to impact damage than pristine panels. Similar results have been reported previously in other works. For example, Ref. [29] reported that the addition of nylon nanofibers to the interfaces of carbon fiber epoxy composites significantly increased the threshold impact force (the force to cause initiation of impact damage) up to 60%. Additionally, this study shows that the impact damage area decreases considerably due to the interleaving with nylon nanofibers. Other works, such as [30], indicated that the incorporation of a mixture of polycaprolactone and nylon nanofibers into the composite ply interfaces could decrease the impact damage area by up to 59.3% in glass fiber epoxy composites. Others authors, such as [31,32], suggest that the interleaving with other electrospun nanofibers such as polyvinylidene fluoride and polyacrylonitrile is not a good choice for toughening epoxy and improving the impact damage resistance of glass fiber/epoxy laminates. In conclusion, it can be said that the interleaving with polycaprolactone nanofibers decreases the impact damage area in glass fiber epoxy composites by more than 20%. This discovery is of the utmost importance for structures made of composite materials (e.g., aircrafts or wind turbines), where the impacts caused by bird strikers or hailstorms are a main concern.

6. Conclusions

This research investigates some important applied properties of polycaprolactone modified glass fiber epoxy composites. It deals with their dynamic properties and their impact resistance. The performed experimental investigations suggest that the incorporation of polycaprolactone nanofibers into composite interfaces increases the damping ratio and does not affect to the natural frequencies. This is important from the view point of using such materials as part of any structural elements that experience vibration. The interleaving with polycaprolactone nanofibers is potentially capable of reducing the amplitude of such vibration.

The impact resistance of materials is very important in a number of industries, including aircraft design and production. The numerical investigation offered in this research shows that the incorporation of polycaprolactone nanofibers can improve the impact resistance of glass fiber epoxy layered composites. Thus, polycaprolactone nanofibers can be used to develop composite materials with improved impact resistance which can be a used for an important number of applications, including aircrafts and wind turbines.

This work presents solid progress toward the practical applications of composites reinforced with polycaprolactone nanofibers as per example the control of vibrations and prevention of damage from delamination.

Author Contributions: C.G. designed, wrote, and revised the paper. I.T. supervised the research and revised the paper and A.Z. provided the composite specimens.

Funding: This research received no external funding.

Acknowledgments: The authors wish to specially thank the University of Bologna for their technical assistance and valuable support in terms of materials and equipment.

Conflicts of Interest: The authors declare no conflict of interest.

Appendix A

Table A1. Material Constants used to simulate the layers of unidirectional glass fiber epoxy composite supplied by ANSYS Workbench Engineering Data.

Constant	Value
Young's Modulus X direction	45 GPa
Young's Modulus Y direction	10 GPa
Young's Modulus Z direction	10 GPa
Poisson's Ratio XY	0.3
Poisson's Ratio YZ	0.4
Poisson's Ratio XZ	0.3
Shear Modulus XY	5 GPa
Shear Modulus YZ	3.8 GPa
Shear Modulus XZ	5 GPa

Appendix B

Since the maximum traction (T_{max}), displacement at debonding (δ_c), and ratio of δ^* and δ_c (α) cannot be determined by experimental tests, these values are determined through a comparison of the experimental results with the numerical simulations of the same tests, allowing for an estimation of the unknown material properties for the cohesive zone model. By definition, the area under the triangle (see Figure 3) corresponds to the critical interlaminar fracture energy for the glass fiber composites. Therefore, the numerical critical fracture energy for the pristine and nano composites is 675 J/m^2 and 490 J/m^2, respectively. These results are in the range of other composite laminates with similar characteristics.

References

1. Daelemans, L.; Vander, H.S.; Baere, I.; Rahier, H.; Paepegem, W.; Clerk, K. Damage resistant composites using electrospun nanofibers: A multiscale analysis of the toughening mechanisms. *ACS Appl. Mater. Interfaces* **2016**, *8*, 11806–11818. [CrossRef] [PubMed]
2. Shinde, D.K.; Kelkar, A.D. Effect of TEOS electrospun nanofiber modified resin on interlaminar shear strength of glass fibre/epoxy composite. *Int. J. Mater. Metall. Eng.* **2014**, *8*, 54–59.
3. Xiao, L.; Xu, L.; Yang, Y.; Zhang, S.; Huang, Y.; Bielawsky, C.W.; Geng, J. Core-shell structured polyamide 66 nanofibers with enhanced flame retardancy. *ACS Omega* **2017**, *2*, 2665–2671. [CrossRef]
4. Palazzetti, R.; Zucchelli, A. Electrospun nanofibers as reinforcement for composite laminates materials—A review. *Compos. Struct.* **2017**, *182*, 711–727. [CrossRef]
5. Jiang, S.; Chen, Y.; Duan, G.; Mei, C.; Greiner, A.; Agarwal, S. Electrospun nanofiber reinforced composites: A review. *Polym. Chem.* **2018**, *9*, 2685–2720. [CrossRef]

6. Garcia, D.; Palazzetti, R.; Trendafilova, I.; Fiorini, C.; Zucchelli, A. Vibration based delamination diagnosis and modelling for composite laminate plates. *Compos. Struct.* **2015**, *130*, 155–162. [CrossRef]
7. Garcia, D.; Tcherniak, D.; Trendafilova, I. Damage Assessment for wind turbine blades based on a multivariate statistical approach. *J. Phys. Conf. Ser.* **2015**, *1*, 628. [CrossRef]
8. Garcia, C.; Wilson, J.; Trendafilova, I.; Yang, L. Vibratory behaviour of glass fibre reinforced polymer (GFRP) interleaved with nylon nanofibers. *Compos. Struct.* **2017**, *176*, 923–932. [CrossRef]
9. Palazzetti, R.; Zucchelli, A.; Trendafilova, I. The self-reinforcing effect of nylon 6,6 nano-fibres on CFRP laminates subjected to low-velocity impact. *Compos. Struct.* **2013**, *106*, 661–671. [CrossRef]
10. Saghafi, H.; Palazzetti, R.; Zucchelli, A.; Minak, G. Influence of electrospun nanofibers on the interlaminar properties of unidirectional epoxy resin/glass fiber composite laminates. *J. Reinf. Plast. Compos.* **2015**, *34*, 907–914. [CrossRef]
11. Palazzetti, R. Flexural behavior of carbon and glass fiber composite laminates reinforced with nylon 6,6 electrospun nanofibers. *J. Compos. Mater.* **2015**, *49*, 3407–3413. [CrossRef]
12. Beylergil, B.; Tanoglu, M.; Aktaş, E. Enhancement of interlaminar fracture toughness of carbon fiber-epoxy composites using polyamide-6,6 electrospun nanofibers. *J. Appl. Polym. Sci.* **2017**, *10*, 45244. [CrossRef]
13. Beckermann, G.W.; Pickering, K.L. Mode I and mode II interlaminar fracture toughness of composite laminates interleaved with electrospun nanofiber veils. *Compos. Part A* **2015**, *72*, 11–21. [CrossRef]
14. Molnar, K.; Kostakova, E.; Meszaros, L. The effect of needleless electrospun nanofibrous interleaves on mechanical properties of carbon fabrics/epoxy laminates. *Express Polym. Lett.* **2014**, *8*, 62–72. [CrossRef]
15. Manh, C.V.; Choi, H.J. Enhancement of interlaminar fracture toughness of carbon fiber/epoxy composites using silk fibroin electrospun nanofibers. *Polym. Plast. Technol. Eng.* **2016**, *55*, 1048–1056. [CrossRef]
16. Akangah, P.; Shivakumar, K. Impact damage resistance and tolerance of polymer nanofiber interleaved composite laminates. In Proceedings of the 53rd AIAA/ASME/ASCE/AHS/ASC Structures, Structural Dynamics and Materials Conference, Honolulu, HI, USA, 23–26 April 2012.
17. Saghafi, H.; Brugo, T.; Minak, G.; Zucchelli, A. Improvement the impact damage resistance of composite materials by interleaving polycaprolactone nanofibers. *Eng. Solid Mech.* **2015**, *3*, 21–26. [CrossRef]
18. Schueren, L.V.; Schoenmaker, B.; Kalaoglu, O.I.; Clerk, K. An alternative solvent system for the steady state electrospinning of polycaprolactone. *Eur. Polym. J.* **2011**, *47*, 1256–1263. [CrossRef]
19. Rahman, N. An efficient method for frequency calculation of an audio signal. *Presidency* **2013**, *2*, 41–45.
20. Casiano, M.J. *Extracting Damping Ratio from Dynamic Data and Numerical Solutions*; Marshal Space Flight Center: Huntsville, AL, USA, 2016; p. 2.
21. Alfano, G.; Crisfield, M.A. Finite element interface models for the delamination analysis of laminated composites: Mechanical and computational issues. *Int. J. Numer. Methods Eng.* **2001**, *50*, 1701–1736. [CrossRef]
22. Deuschle, H.M.; Puck, A. Application of the Puck failure theory for fibre-reinforced composites under three-dimensional stress: Comparison with experimental results. *J. Compos. Mater.* **2012**, *47*, 827–846. [CrossRef]
23. Garcia, C.; Trendafilova, I.; Zucchelli, A.; Contreras, J. The effect of nylon nanofibers on the dynamic behaviour and the delamination resistance of GFRP composites. In Proceedings of the International Conference on Engineering Vibration ICoEV, MATEC Web Conference, Sofia, Bulgaria, 4–7 September 2017; Volume 148, p. 14001.
24. Gou, J.; O'Braint, S.; Gu, H.; Song, G. Damping augmentation of nanocomposites using carbon nanofiber paper. *J. Nanomater.* **2006**, *2006*, 32803. [CrossRef]
25. Her, S.H.; Lai, C.Y. Dynamic behaviour of nanocomposites reinforced with multi-walled carbon nanotubes (MWCNTs). *Materials* **2013**, *6*, 2274–2284. [CrossRef] [PubMed]
26. Padal, K.T.B.; Ramji, K.; Prasad, V.V.S. Damping behaviour of jute nano fibre reinforced composites. *Int. J. Emerg. Technol. Adv. Eng.* **2014**, *4*, 753–759.
27. Kordani, N.; Fereidoon, A.; Ashoori, M. Damping augmentation of nanocomposites using carbon nanotube/epoxy. *J. Struct. Dyn.* **2011**, *3*, 1605–1615.
28. Zhang, J.; Yang, T.; Lin, T.; Wang, C.H. Phase morphology of nanofiber interlayers: Critical factor for toughening carbon/epoxy composites. *Compos. Sci. Technol.* **2012**, *72*, 256–262. [CrossRef]
29. Akangah, P.; Lingaiah, S.; Shivakumar, K. Effect of nylon-66 nano-fiber interleaving on impact damage resistance of epoxy/carbon fiber composite laminates. *Compos. Struct.* **2010**, *92*, 1432–1439. [CrossRef]

30. Saghafi, H. Mechanical Behaviour of Flat and Curved Laminates Interleaved by Electrospun Nanofibers. Ph.D. Thesis, University of Bologne, Bologne, Italy, 2013.
31. Saghafi, H.; Palazzetti, R.; Zucchelli, A.; Minak, G. Impact response of glass/epoxy laminate interleaved with nanofibrous mats. *Eng. Solid Mech.* **2013**, *1*, 85–90. [CrossRef]
32. Saghafi, H.; Palazzetti, R.; Minak, G.; Zucchelli, A. Effect of PAN nanofiber interleaving on impact damage resistance of GFRP laminates. In Proceedings of the 6th International Conference on Nanomaterials—Research and Application, Brno, Czech Republic, 5–7 November 2014.

© 2018 by the authors. Licensee MDPI, Basel, Switzerland. This article is an open access article distributed under the terms and conditions of the Creative Commons Attribution (CC BY) license (http://creativecommons.org/licenses/by/4.0/).

MDPI

St. Alban-Anlage 66

4052 Basel

Switzerland

Tel. +41 61 683 77 34

Fax +41 61 302 89 18

www.mdpi.com

Journal of Composites Science Editorial Office

E-mail: jcs@mdpi.com

www.mdpi.com/journal/jcs

www.ingramcontent.com/pod-product-compliance
Lightning Source LLC
Chambersburg PA
CBHW051906210326

41597CB00033B/6049